NAPTOWN ROCK RADIO WARS

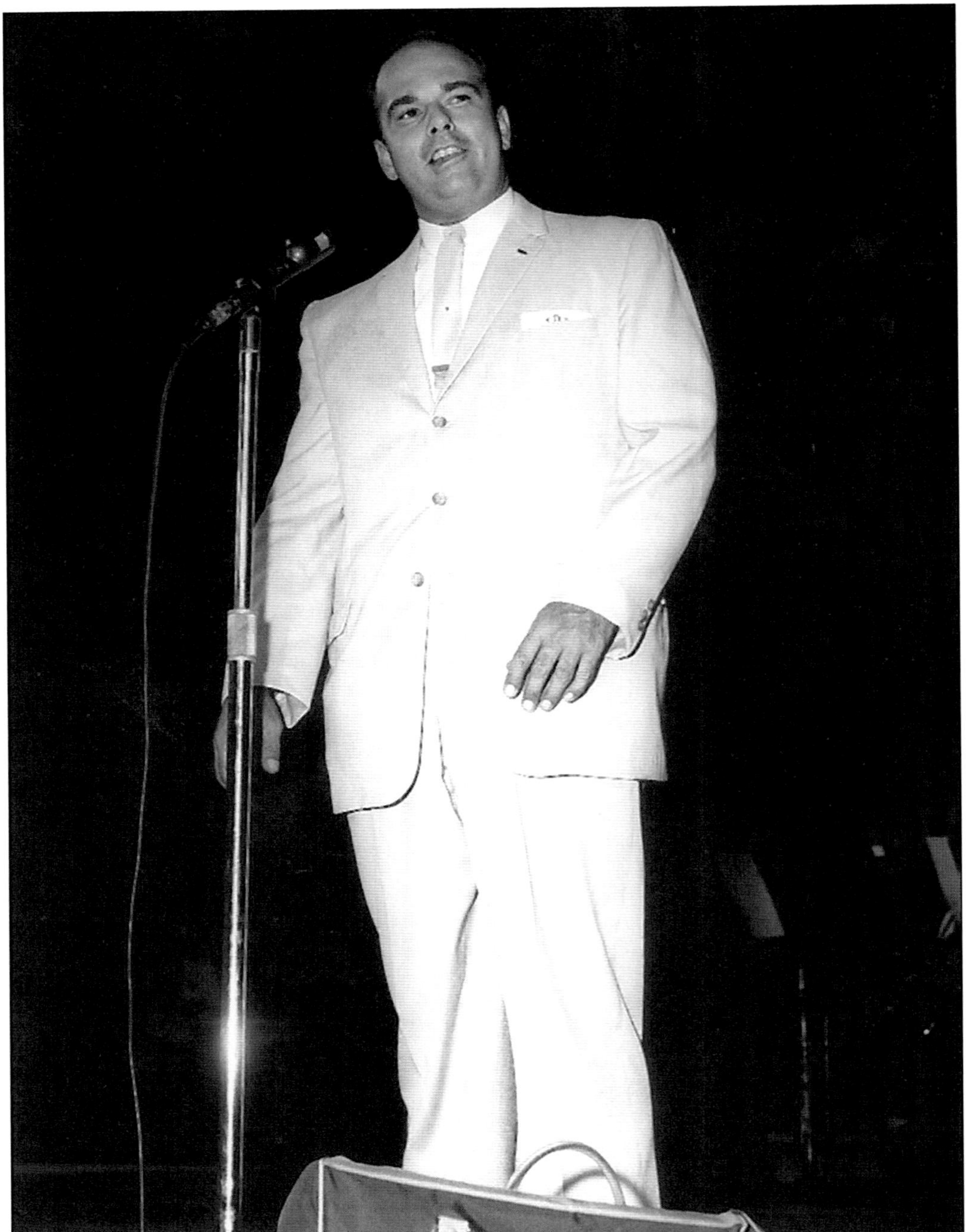

"Bouncin' " Bill Baker was the morning-drive deejay for WIBC-AM in the late 1950s and early 1960s. His wacky mix of character voices and rock 'n' roll music captured a huge and unprecedented audience in Indianapolis. Baker is largely considered to have created "Ground Zero" for rock radio in central Indiana.

ON THE COVER: The iconic RCA 77DX ribbon microphone was the standard of recording studios and broadcasting throughout the 1950s and 1960s. This particular one is owned by Milo Miller and was photographed by Spark Art Photography in Indianapolis.

NAPTOWN ROCK RADIO WARS

David Fulton

ARCADIA
PUBLISHING

ISBN 978-0-7385-9851-2

Published by Arcadia Publishing
Charleston, South Carolina

Printed in the United States of America

Library of Congress Control Number: 2012946318

For all general information, please contact Arcadia Publishing:
Telephone 843-853-2070
Fax 843-853-0044
E-mail sales@arcadiapublishing.com
For customer service and orders:
Toll-Free 1-888-313-2665

Visit us on the Internet at www.arcadiapublishing.com

This book is dedicated to my sons, Rob Fulton and Chris Fulton, and my film's producers, Al Stone and Brad Schuchard.

Contents

Acknowledgments

Throughout the book, quotes are provided by:

Peggy McDowell	Radio Personality
"Big" John Gillis	WNAP Disc Jockey
Bill Shirk	WXLW Owner/Disc Jockey
Art Vuolo Jr.	"Radio's Best Friend"
Pat Kelly	WIFE Sales Manager
Al Stone	WNAP Program Director
James McDowell	WIRE Disc Jockey
Jim Hilliard	Fairbanks Broadcasting, President
George Johns	Fairbanks Broadcasting, National Program Director
Reb Porter	WIFE Disc Jockey
Bobby Bernard	WIFE Salesperson
"Steve Cooper"	WIFE Disc Jockey
Cris Conner	WNAP Disc Jockey
Ann Craig-Cinnamon	WNAP Disc Jockey
Mike Griffin	WNAP Disc Jockey
Bill Donnella	WIFE News Director
Dave "Mike O'Brien" Heck	WIFE Disc Jockey
Don Worsham	Radio Free Naptown Disc Jockey
Jack Marsella	WIFE Vice President of Sales
Jeff Smulyan	CEO Emmis Communications
Steve "Freddie Fever" Reynolds	WNAP Disc Jockey
"Shotgun Tom" Kelly	KCBQ Disc Jockey
Roger W. Morgan	WIFE Disc Jockey
Tom Cochrun	WNAP Newsman
Scott Shannon	WPIJ NYC Disc Jockey
Scott Wheeler	WIFE Disc Jockey
Bob Kiley	WIFE General Manager
Michael "Buster Bodine" Hanks	WNAP Disc Jockey
Neil Bolding	Radio Free Naptown Disc Jockey
Tom Woody	Radio Free Naptown Disc Jockey
Al "The Bishop" Hobbs	WTLC General Manager

Images published courtesy of Videopolis.tv

Special thanks to Jed Duvall, Alan Warner, Bob Moore, John Rabold, Rick Wilkerson/Keywerx, Jon Quick/absolutemaxpr.com, Susan DeVine, Milo Miller, Joy Carter, Gina King, and Loughmiller's Restaurant.

Introduction

Trends found in decades do not really break neatly at the count of 10. What we recall as the 1950s really stretched well into the early 1960s, and those wacky late 1960s have never really ended for some. They gave way to a cultural revolution on such a myriad of levels that they all are still yet to be counted and accounted for. It was the time when the majority demographic of postwar American youth, what would later be called "baby boomers," shook itself free, stood up, and shouted "Rock and Roll!" This concept would be presented, extolled, damned, and exploited.

WIFE came to Indianapolis, also dubbed "Naptown," in 1963 like an aerial bomb in a sacred moment of silence. Concurrent with an assassinated JFK and the Beatles, WIFE Radio came to town with hot jingles, the white-hatted interchangeable "Good Guys," a lot of contests, and endless promotions, and it happened swiftly. Records were less than three minutes, and jocks shut up after 30 seconds, abiding by the concept of "say it, play it, play the jingle, pump the contest." Indeed, Top 40 had come to town. WIFE's signal was wimpy, but its voice was loud. Employees knew the term "flop sweat" from the pressure inside the studio and its "Window to the World." WIFE printed money at a staggering pace. Don Burden drove this station and his other radio machines "pedal to the metal" until he finally went off the Federal Communications Commission cliff. That journey took some time—and it was a hell of a trip.

Meanwhile, up the way at the beautiful "Two-Story Brick" on North Illinois Street, heads were spinning. WIBC had been on top in Indianapolis radio for years. It was knocked off its axis in what seemed like less time than it will take to read this introduction. WIFE had delivered a knockout punch like that Clay kid down in Louisville. WIBC had, without dispute, one of the finest, most relied upon newsrooms in the country under the iron direction of Fred Heckman. They were not the Good Guys, they were the disaster go-to guys. The Hoosier mindset had always been if a person saw smoke, it was time to tune to WIBC. Yet suddenly the city was "spending night and day with your WIFE," or so said the billboard out by the airport.

Somebody pulled the plug. Somebody forgot to start the tape. Somebody screwed up, and WIBC's spot on the FM dial all but disappeared—nobody called. Outside of the building, no one cared. It is the fear of every fledgling deejay: there really was no one out there. It was 1968, and those at "the mother ship" figured if no one is listening, they might as well mess with WIFE.

The crowded dials of the "Big Three" cities—New York, Chicago, and Los Angeles—had already ventured into the unknown swamps of FM, so why not Naptown? Hiring tireless, creative young announcers at bargain-basement salaries and embracing a venue few had direct access to, WIBC-FM was suddenly no more, and WNAP was born. To the strains of "In-A-Gadda-Da-Vida" came the personalities King Freak and Michael T, John G, Big John, Booger, Buster, Major Tom Lewis, your friend for life "Mr. Below," the Smash, and Rex Rona. WNAP set the beat and stole WIFE's lunch with the illusion of underground radio, the 25th Hour, constant concerts, and the original morning team. Theirs was the Fantasy Park, the

Free Fair, the WNAP Raft Races, and eventually that smiling Buzzard who would devour the hapless Good Guys.

Dave Fulton has told the story, and he has told it well. He has tracked down the glimpses, the bites, the archives, and most of all the survivors—at least those willing to talk. And from these separate strings, threads, and skeins, he has woven a fascinating tale. When can you see it? Like my friends down in Jamaica say, "Soon come."

—Allen Deck, retired
Department of Communications
Butler University

One

The Dawn of Rock 'n' Roll Radio

Radio lost all of its most popular programming to television in the late 1940s and early 1950s. All of the soap operas, quiz shows, detective dramas, and comedies were stolen right out from under radio by the new upstart medium where a person could see what was going on as well as hear. With the big shows migrating to television, a vacuum was left in radio.

In addition to television, there were other lasting innovations coming into the marketplace. The world saw one if its first format wars when Columbia Records came out with a new concept that would replace 78-rpm records: the LP, or "long playing" record. It was perfect for longer musical works. Radio Corporation of America's (RCA) solution to 78s was the 45-rpm record, which was smaller and much less fragile than a 78. The 45 was great for selling the hits.

At the time, AM radio was king, but FM had come out of engineers' labs and was starting to find listeners. The sound on FM was unquestionably purer: much better fidelity and less prone to static. But the party was on AM radio, so why bother?

Then, in the 1950s, all hell broke loose in the music world. Black music and white music crashed together to form what would become rock 'n' roll. Parents were counseled that it was just a passing fad, but the kids would have none of it. The newborn genre had staying power.

Radio stations, desperate to find something to replace the programming stolen by their counterparts in television, turned to a new concept: the disc jockey. Even with the introduction of rock 'n' roll, music was all over the map with classical, big band, easy listening, and vocalist options. Deejays had quite the selection to choose from.

The deejay was the strongest voice on the air. The deejay actually made the station profit, because he was the guy, or she was the girl, who played the record, did the commercial, and brought the listeners in through their personalities. —James McDowell

In the late 1950s, Bouncin' Bill Baker owned radio in Indianapolis. He was the morning man at WIBC, and that was owned by Fairbanks Broadcasting. And the other stations just couldn't compete. You know, we're talking about WIRE, WXLW, WISH, WFBM. Bouncin' Bill just owned the market. —Al Stone

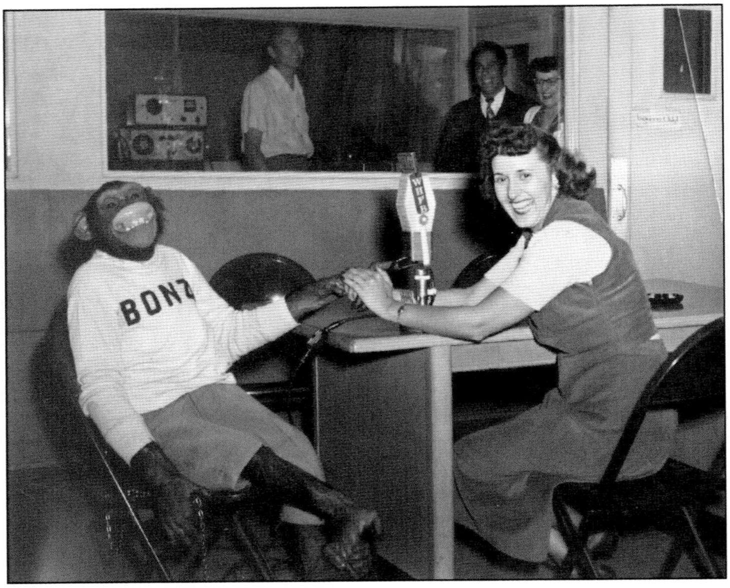

In the 1940s and 1950s, Peggy McDowell (at right, seated) was a producer of radio programming. As a student at Butler University, she learned to work with voice talent, sound-effects technicians, and music.

Radio was fun! And I know they put that across to the people, too. I produced several shows and we would have to have our sound effects. And a lot of them were just handmade. And so everyone listening at home thought that's all happening, you know. They'd laugh if they walked in and saw us doing all of these things. —Peggy McDowell

In the Fifties, you didn't have formatted radio. I would play big band music, symphony music, interview-type talk shows. . . . It was supposed to be a radio station for everybody. —Reb Porter

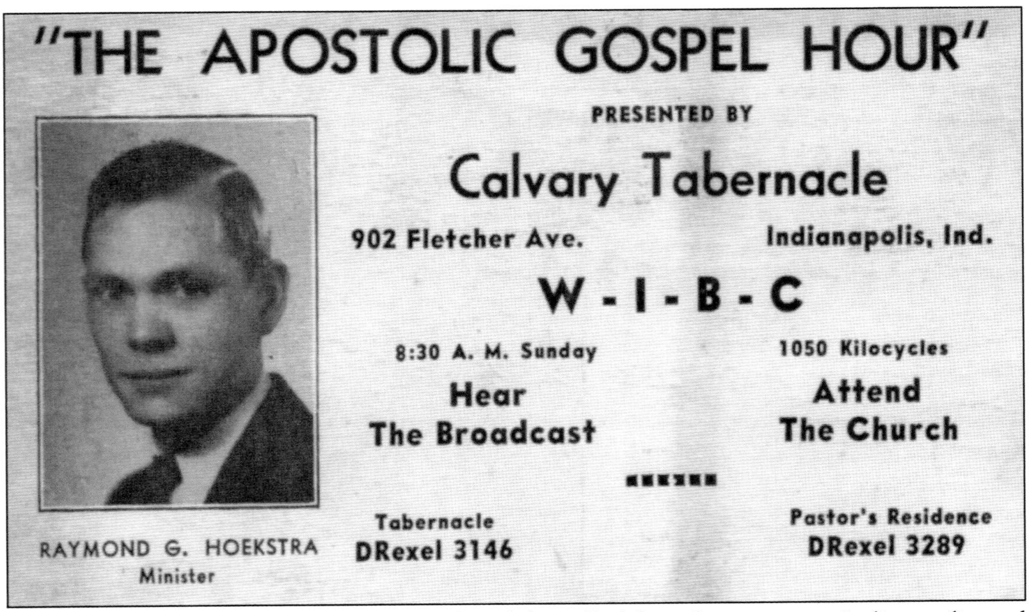
"THE APOSTOLIC GOSPEL HOUR"

PRESENTED BY

Calvary Tabernacle

902 Fletcher Ave. Indianapolis, Ind.

W-I-B-C

8:30 A. M. Sunday 1050 Kilocycles

Hear The Broadcast

Attend The Church

RAYMOND G. HOEKSTRA
Minister

Tabernacle
DRexel 3146

Pastor's Residence
DRexel 3289

At the time, Christian programming was plentiful on all of the radio stations in Indianapolis, and the Sunday schedule would be packed morning and night. Although this postcard invitation is from 1940, the sacred nature of church on Sundays would continue on the radio for decades.

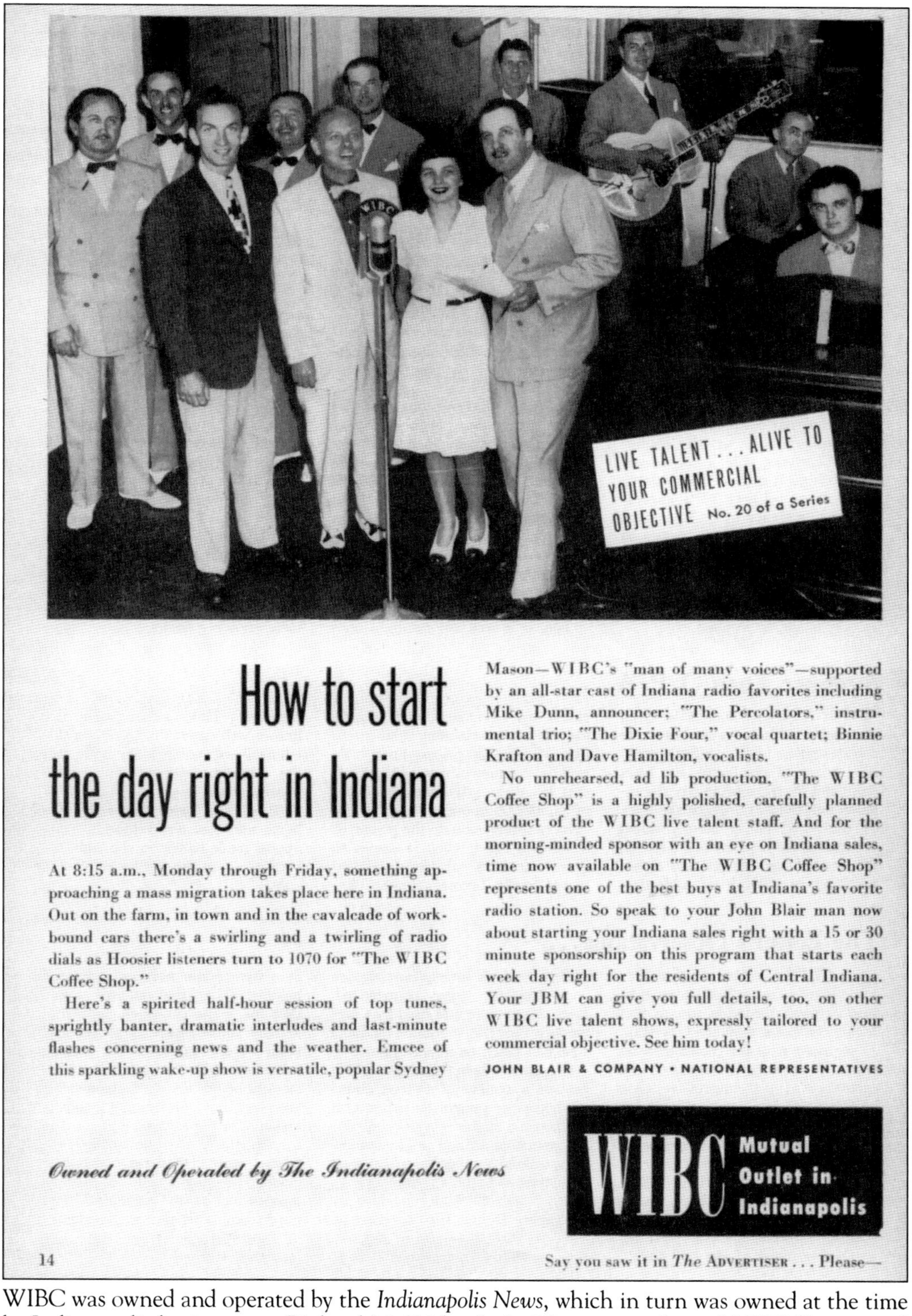

How to start the day right in Indiana

At 8:15 a.m., Monday through Friday, something approaching a mass migration takes place here in Indiana. Out on the farm, in town and in the cavalcade of work-bound cars there's a swirling and a twirling of radio dials as Hoosier listeners turn to 1070 for "The WIBC Coffee Shop."

Here's a spirited half-hour session of top tunes, sprightly banter, dramatic interludes and last-minute flashes concerning news and the weather. Emcee of this sparkling wake-up show is versatile, popular Sydney Mason—WIBC's "man of many voices"—supported by an all-star cast of Indiana radio favorites including Mike Dunn, announcer; "The Percolators," instrumental trio; "The Dixie Four," vocal quartet; Binnie Krafton and Dave Hamilton, vocalists.

No unrehearsed, ad lib production, "The WIBC Coffee Shop" is a highly polished, carefully planned product of the WIBC live talent staff. And for the morning-minded sponsor with an eye on Indiana sales, time now available on "The WIBC Coffee Shop" represents one of the best buys at Indiana's favorite radio station. So speak to your John Blair man now about starting your Indiana sales right with a 15 or 30 minute sponsorship on this program that starts each week day right for the residents of Central Indiana. Your JBM can give you full details, too, on other WIBC live talent shows, expressly tailored to your commercial objective. See him today!

JOHN BLAIR & COMPANY • NATIONAL REPRESENTATIVES

Owned and Operated by The Indianapolis News

WIBC Mutual Outlet in Indianapolis

14 Say you saw it in *The* ADVERTISER . . . Please—

WIBC was owned and operated by the *Indianapolis News*, which in turn was owned at the time by Indianapolis businessman Richard M. Fairbanks Jr. The agency that sold WIBC's national advertising, John Blair and Company, took out a full page ad to lure in sponsors for one of the WIBC morning shows, *The WIBC Coffee Shop*.

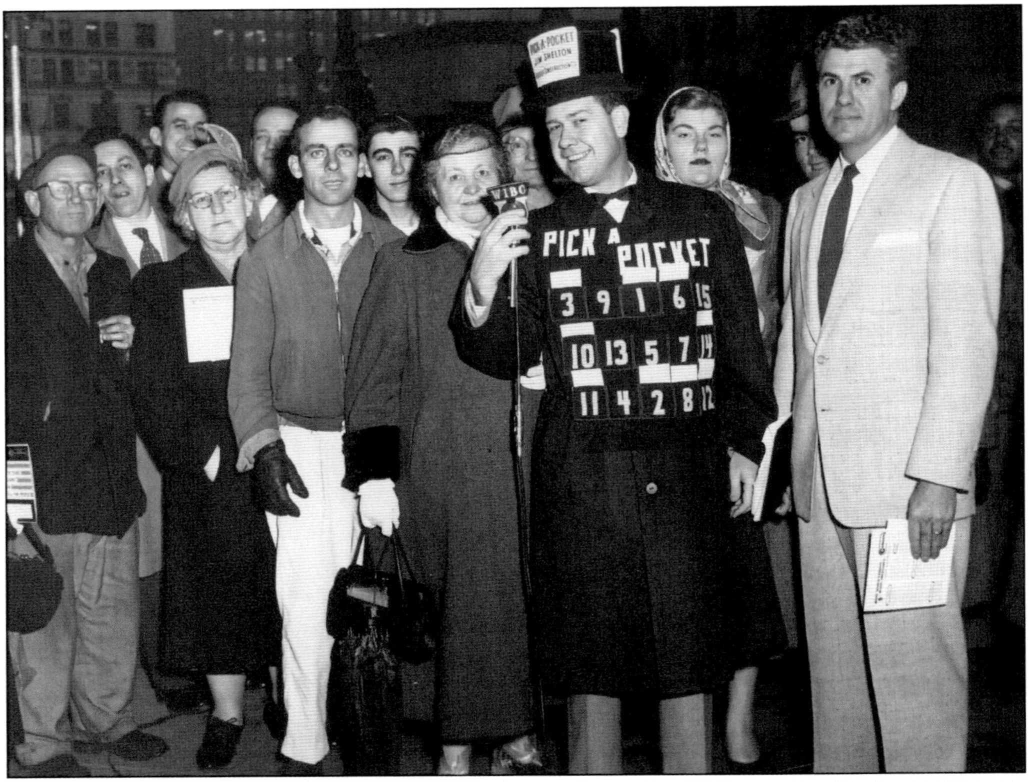

One of the most popular radio segments in Indianapolis in the 1950s was Jim Shelton's "Pick-A-Pocket" contest, a man-on-the-street feature. He had an apron that had several numbered pockets. Each pocket had an item in it, but one pocket contained an item worth up to $50. The object was to guess which pocket had the big prize.

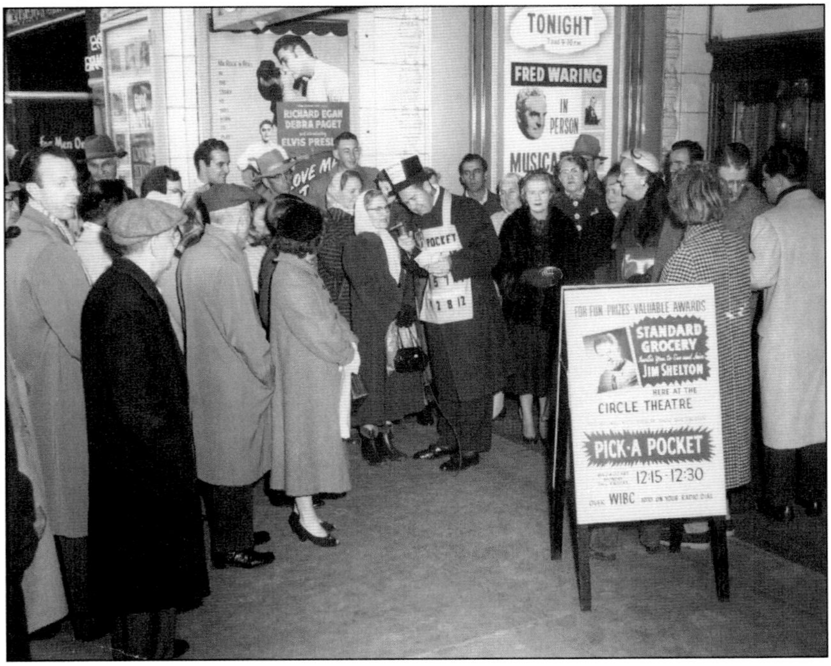

I have a vest that I wear with 15 pockets and they're all mixed up, the numbers on the pockets. And it's the only legal place in town that you can pick a pocket.
—Jim Shelton

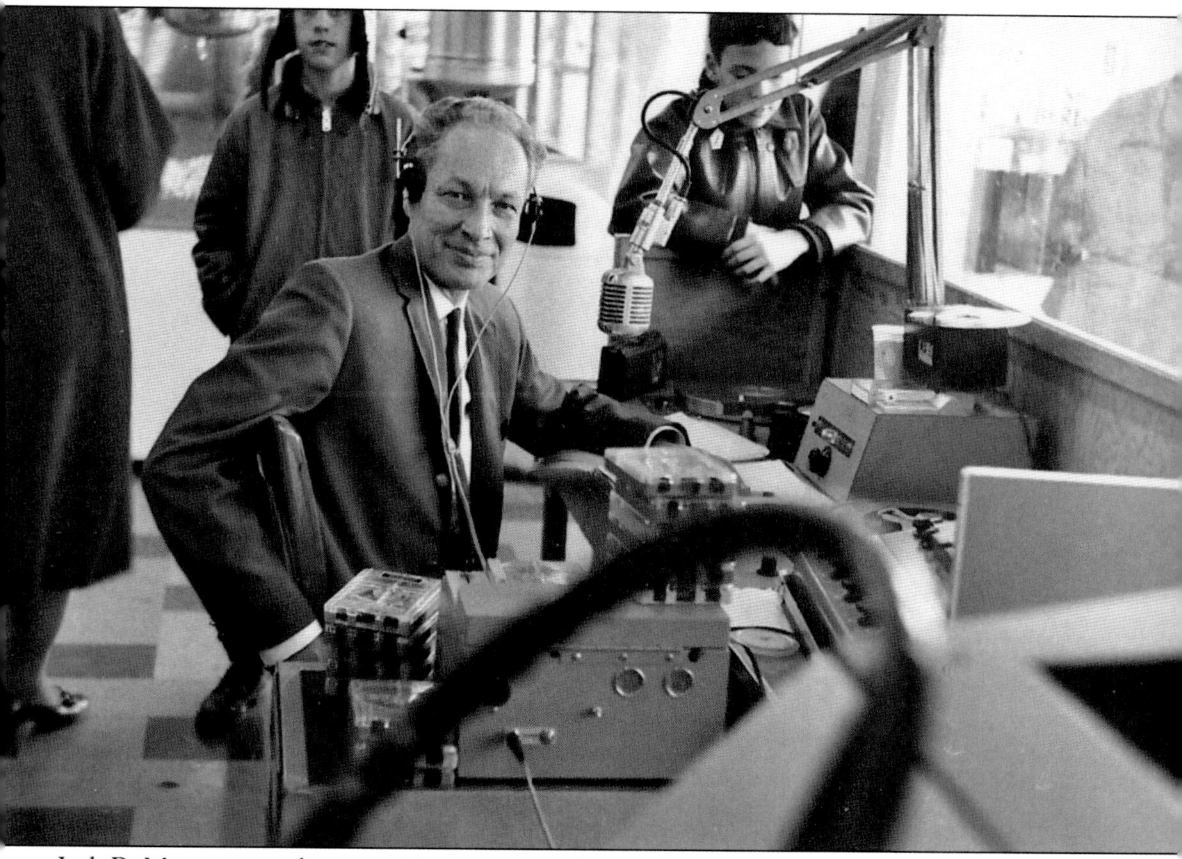

Jack D. Morrow, seen here in 1964, came to Indianapolis in 1940 and was a jack of all trades at WIBC. Among other things, he was an announcer, disc jockey, news reporter, sports color commentator, interviewer, and program director. His friendly, button-down approach to broadcasting was typical of that found on WIBC. Taken in the 1960s, this photograph shows Jack on a remote broadcast, most likely at a suburban Standard supermarket.

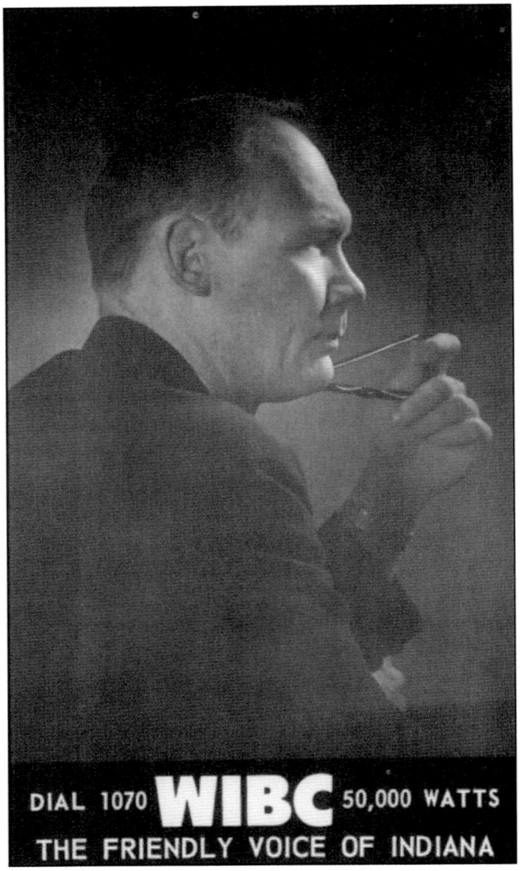

The big push in 1950s radio, more than the music, was the show and the personality. On-air broadcasters would typically become major celebrities in the community. Shown here are Erik James (left) of WIBC and Jerry Chapman (below) of WIRE. James did the overnight broadcast at WIBC and at one point was paired as a sidekick with Bouncin' Bill Baker, with little success.

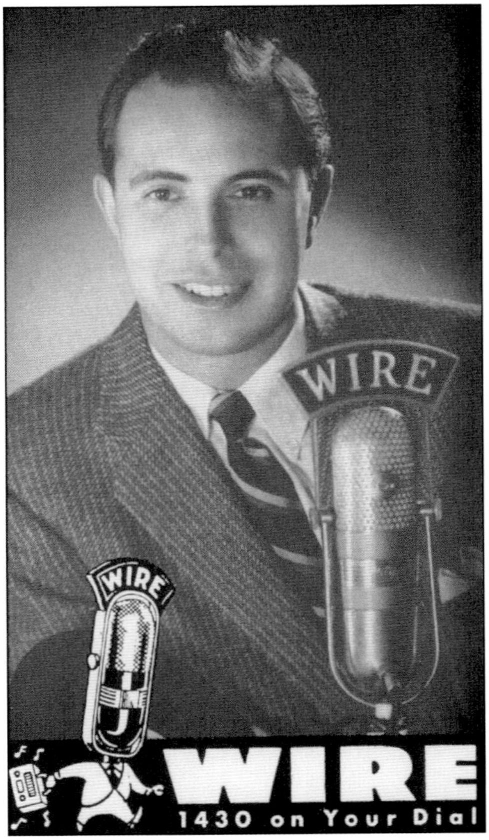

Jerry Chapman's shows were *Record Report* and *High School Hop*. He left WIRE in the early 1960s to go to WFBM. In 1973, he became general manager of WFBM-TV/ WRTV, a position he held for 18 years.

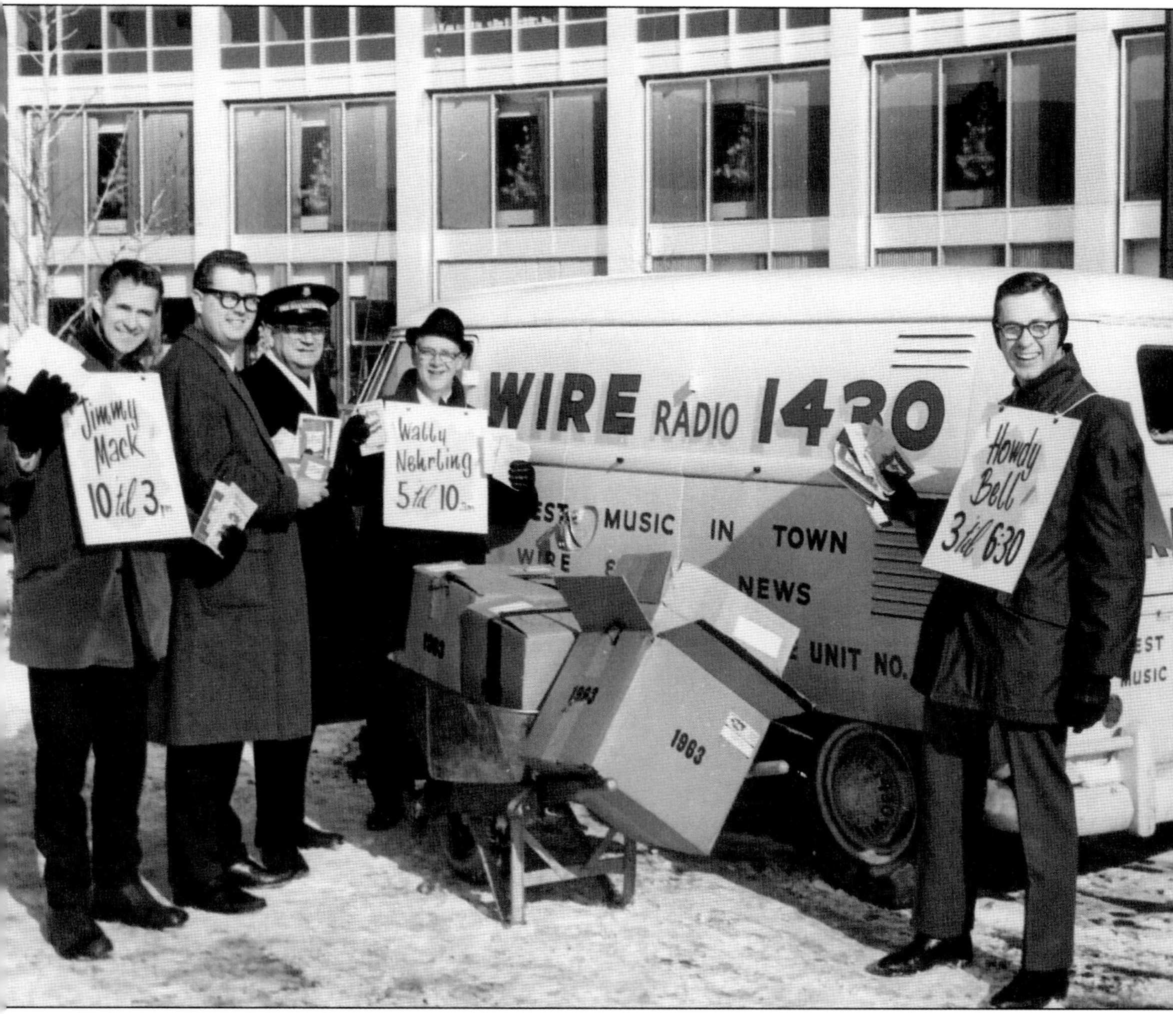

WIRE's format was called "pop standard." The station was owned by the *Indianapolis Star*, and its studios were in the Star Building on North Pennsylvania Street. James "Jimmy Mack" McDowell (left) was a pioneer of youth-oriented radio programming. WIRE broadcast at a frequency of 1430 kilohertz. Pictured with Jimmy Mack are, from left to right, Doug Zinc, a Salvation Army captain, "Uncle" Wally Nehrling, and Howdy Bell.

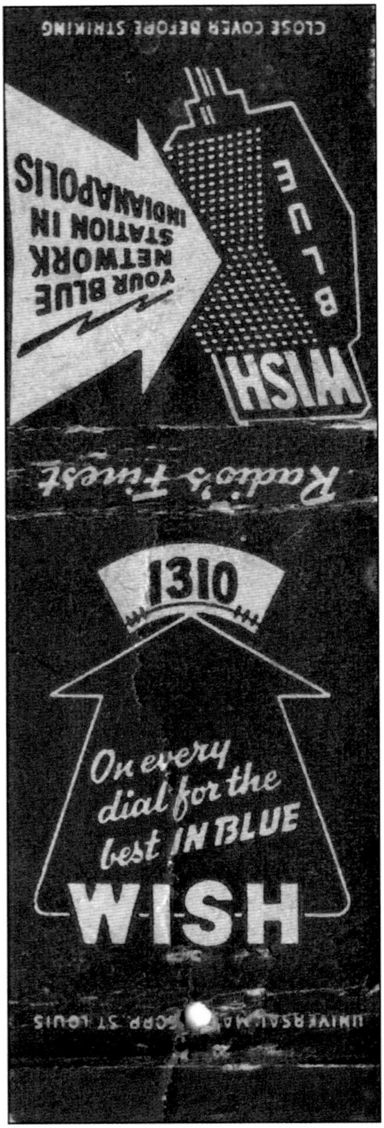

Before 1958, WISH Radio's format was a variety of styles and tastes. On any given day, it would present a little bit of everything. Broadcast at 1310 kilohertz, WISH became a Top 40 station in the late 1950s before transforming into WIFE at the end of 1963.

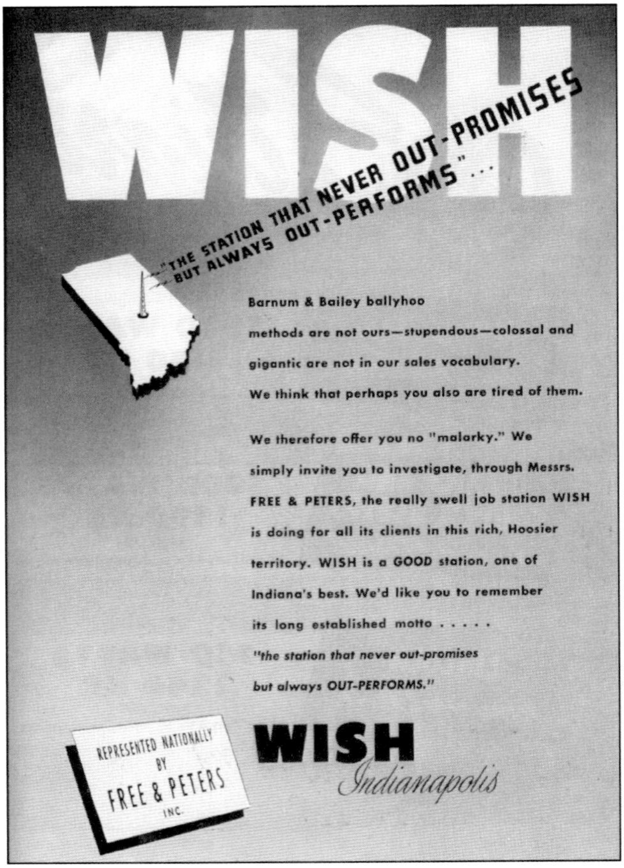

WFBM was the grand dame of radio in Indianapolis at 1260 kilohertz. Licensed in 1924, it was the oldest radio station in Indianapolis. WFBM had the advantage of a sister station on television, which helped in name recognition. Eldon C. Campbell was the general manager for WFBM-AM/FM/TV from 1957 to 1973. Its chief competitor was WIBC.

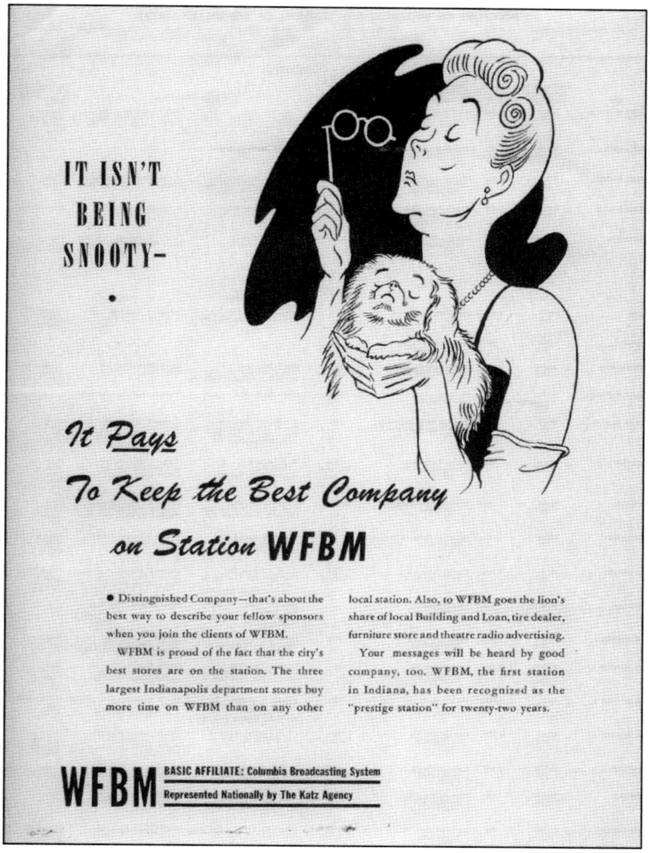

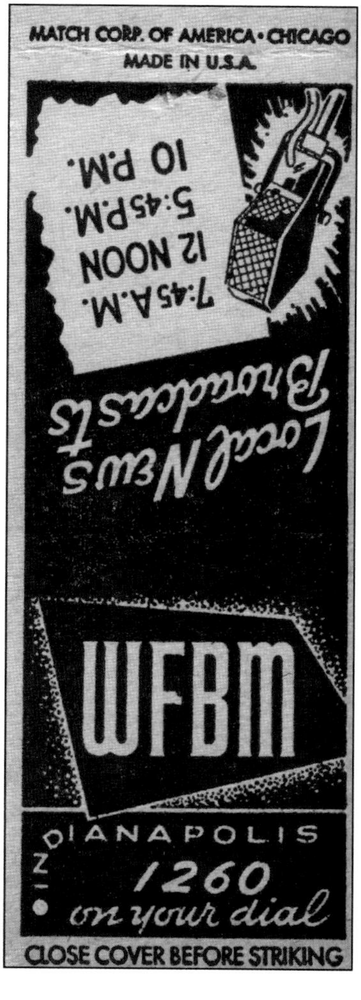

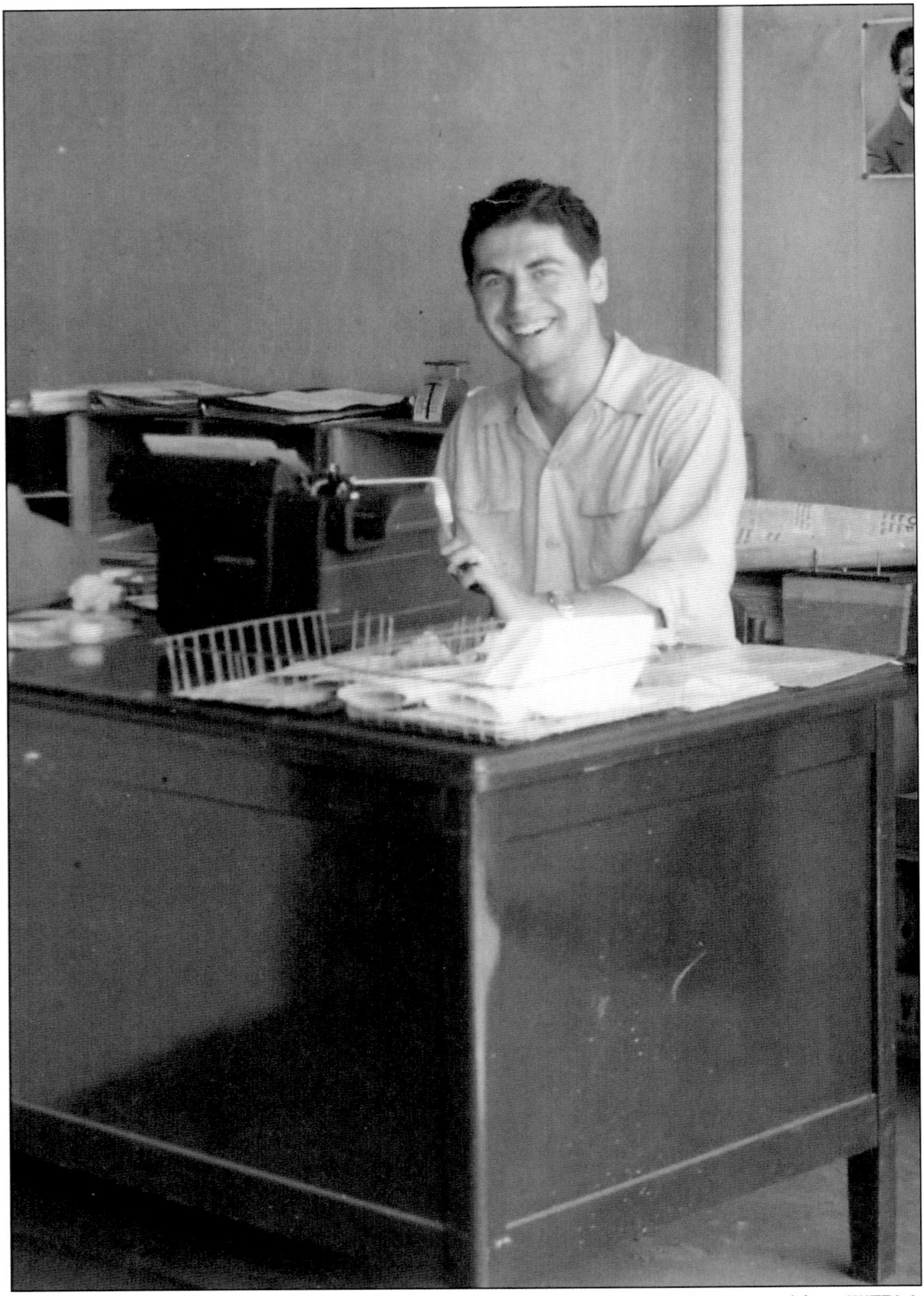

Hal Fryar had two programs on WFBM: *Fryar's Club* and *Hal's a Poppin'*. He later moved from WFBM-AM to WFBM-TV and became a very popular children's show host as "Harlow Hickenlooper."

Merrill's Hi-Decker, along with the Tee-Pee Restaurant, was a popular cruising spot for teens in the 1950s and 1960s. It was located on East Thirty-eighth Street directly across from the Indiana State Fairgrounds. Dick Summer would have an on-air version of "Rate-A-Record," where cars would honk or flash their headlights to show approval.

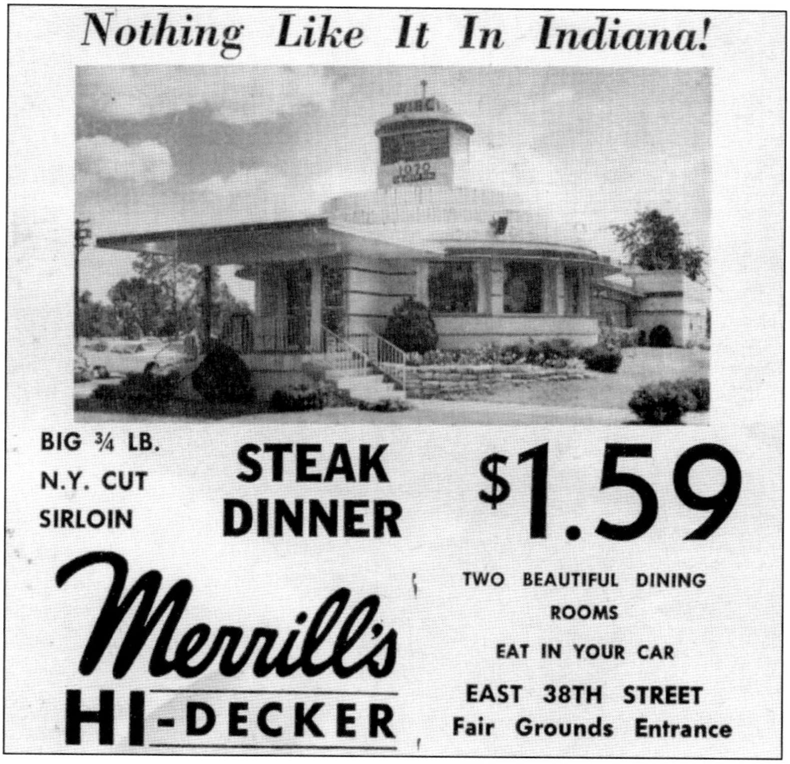

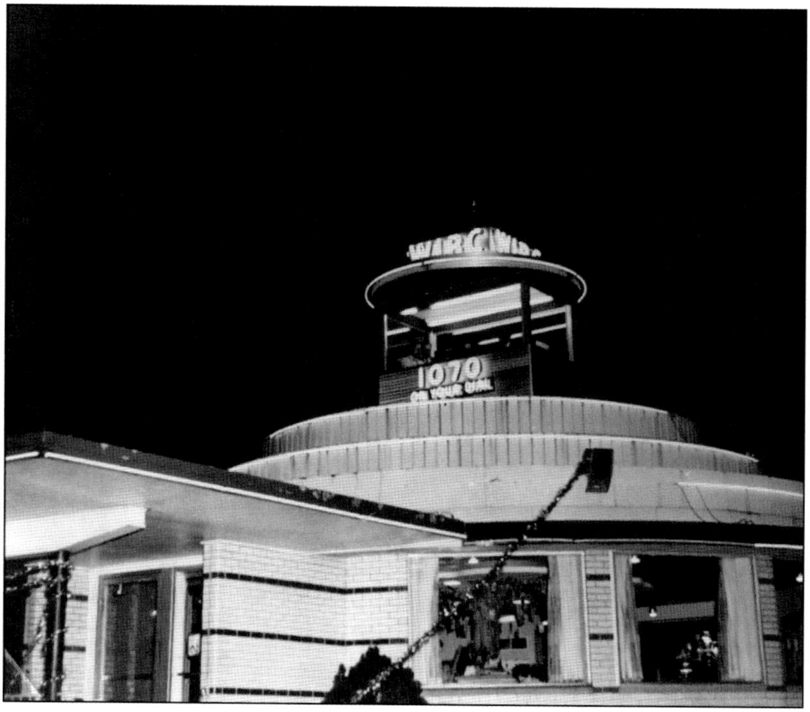

And Dick Summer would do his show from a little studio on top of a drive-in that was reminiscent of Mel's Drive-in in American Graffiti. *—Art Vuolo, Jr.*

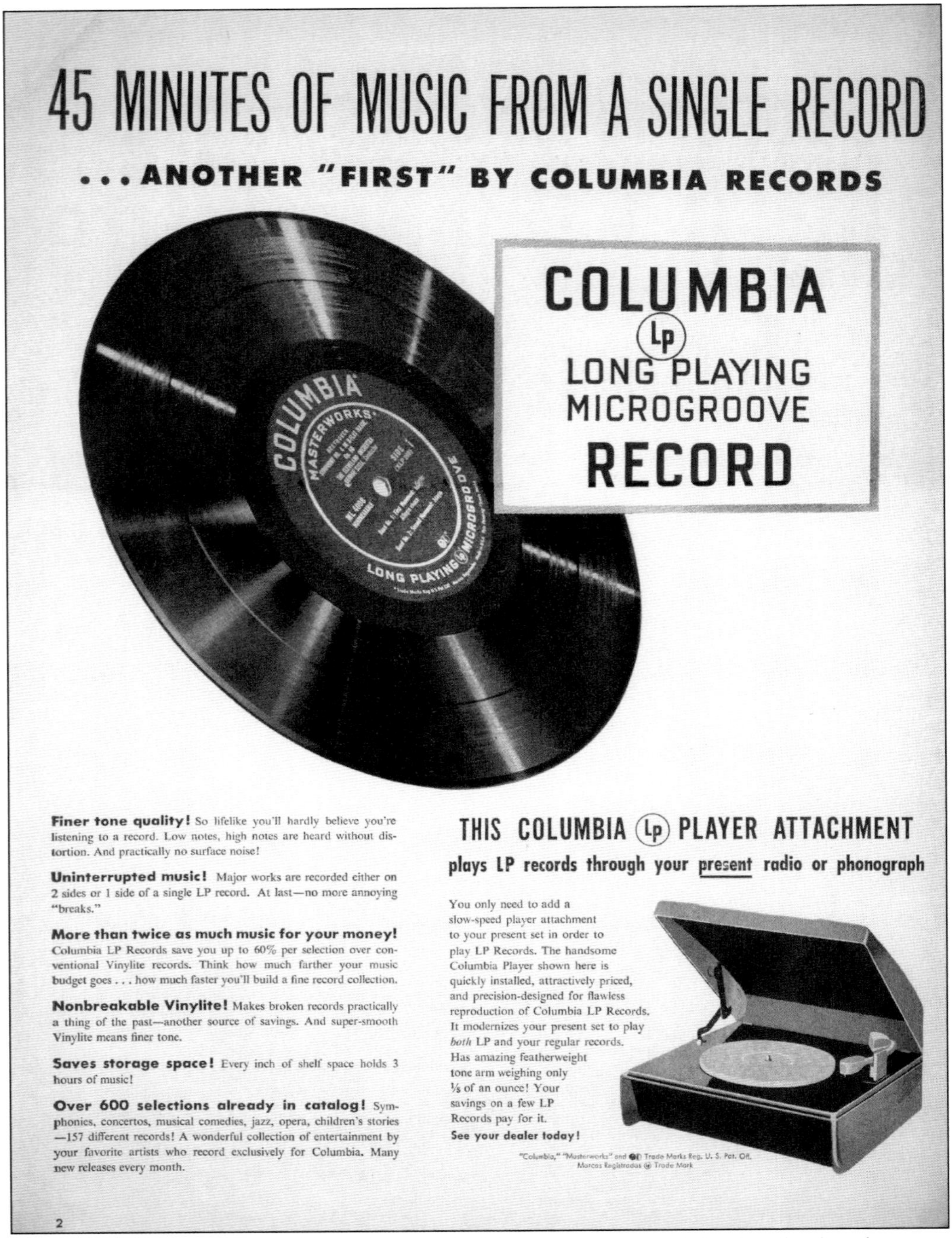

Shellac 78-rpm recordings had been the standard for consumer records for decades, but they were very brittle and not able to hold much music. In the summer of 1948, Columbia Records unveiled a brand-new format designed to replace the 78 called an LP, which used vinyl rather than shellac. Not only did this new technology result in a more durable product that held more music, but the sound quality was substantially improved.

While Columbia Records was revolutionizing the record industry, RCA remained quiet. The company had been secretly developing its own replacement for 78s, but Columbia's LPs were stealing the thunder RCA had been depending on. On January 10, 1949, RCA debuted the 45, so named because the seven-inch disk spun around at 45 rpm. It was the beginning of a format war: the little record with the big hole versus the big record with the little hole. Originally, 45s came in different colors, depending upon the style of music the disk contained.

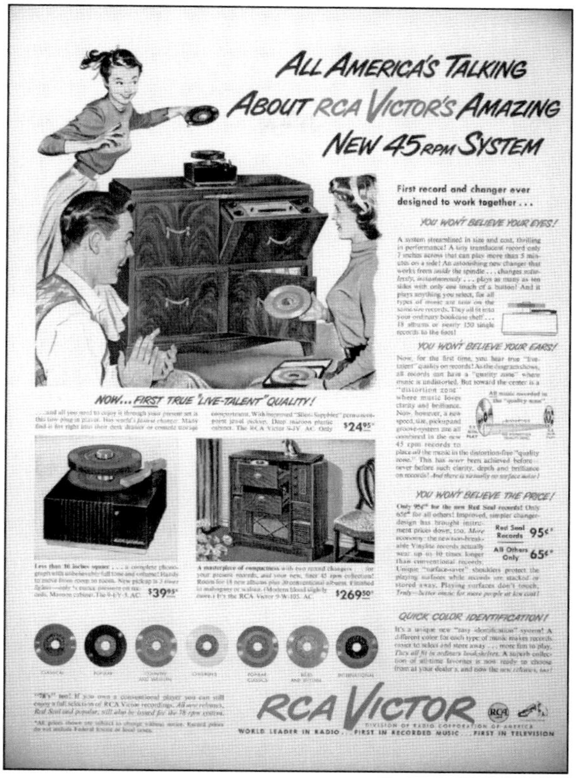

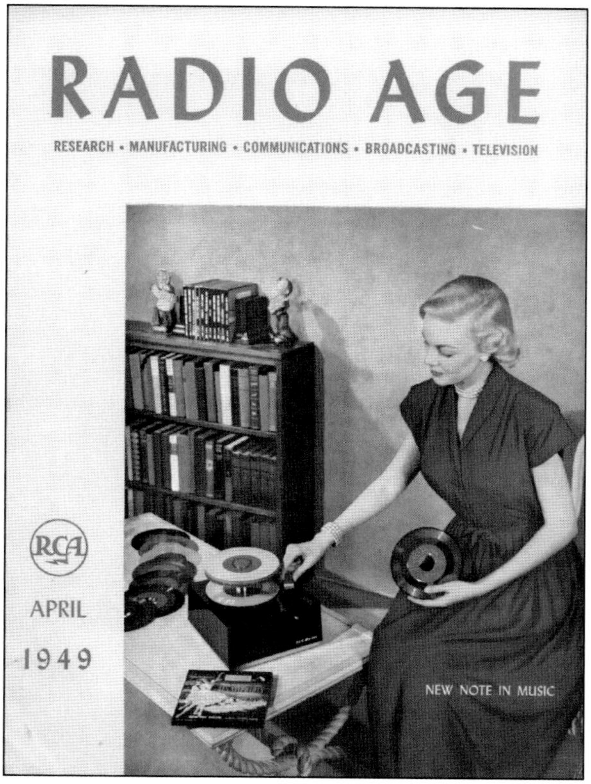

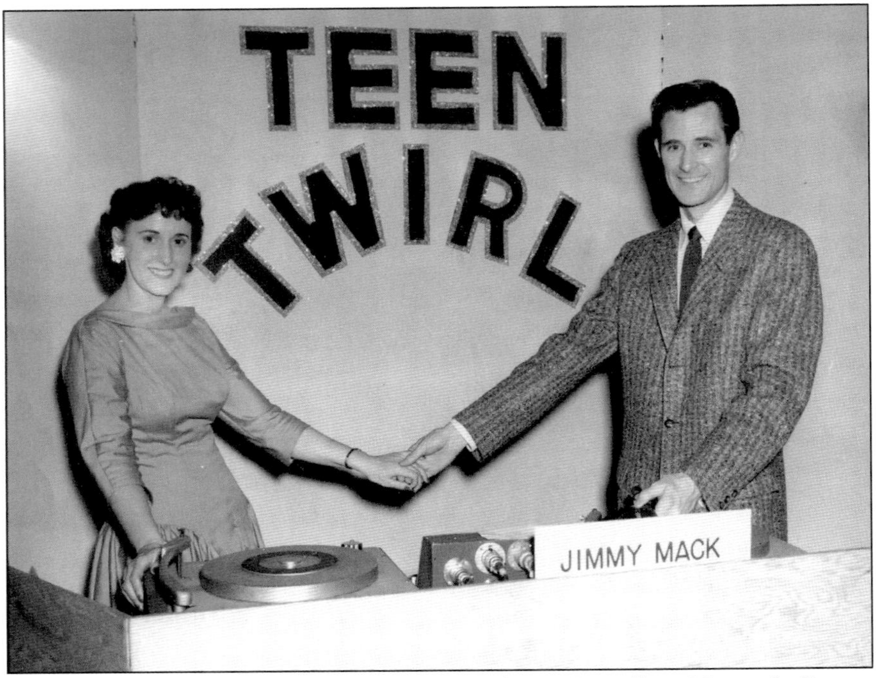

On Saturdays from 1957 to 1962, James "Jimmy Mack" McDowell and his wife, Peggy, hosted a televised sock hop on WISH-TV called *Teen Twirl*, based on Jimmy Mack's WISH-AM radio show, *Teens & Tunes*. He also hosted a daily radio show on WISH-AM called *Melody Mac* and for a year was the local host of a national teen show, *Hi-Fi Club*, sponsored by Coca-Cola. Below, McDowell poses around 1959 with the Deb-Tones, three girls from North Central High School who had won a recording contract with RCA.

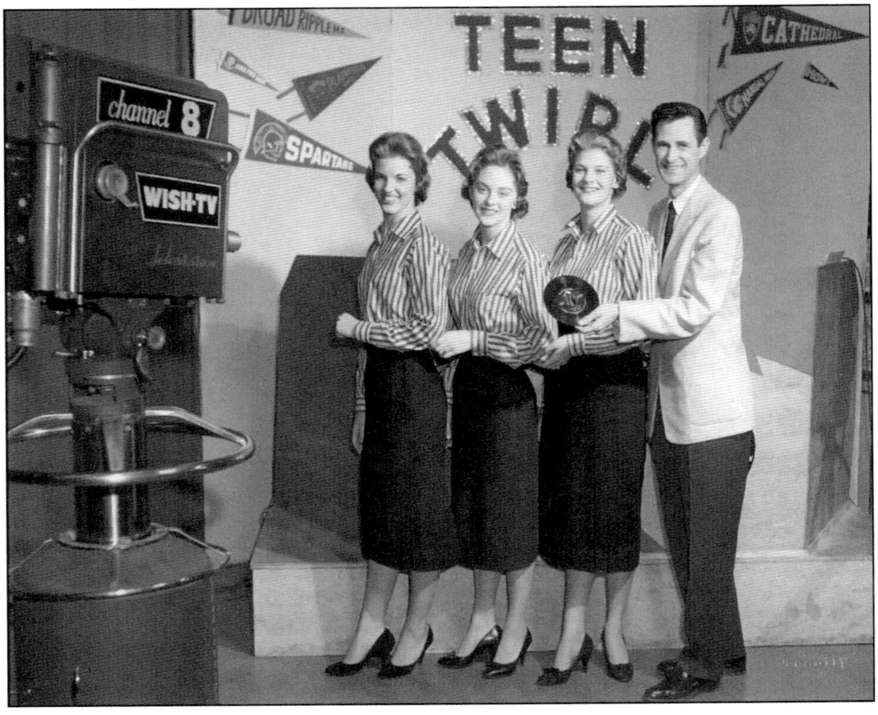

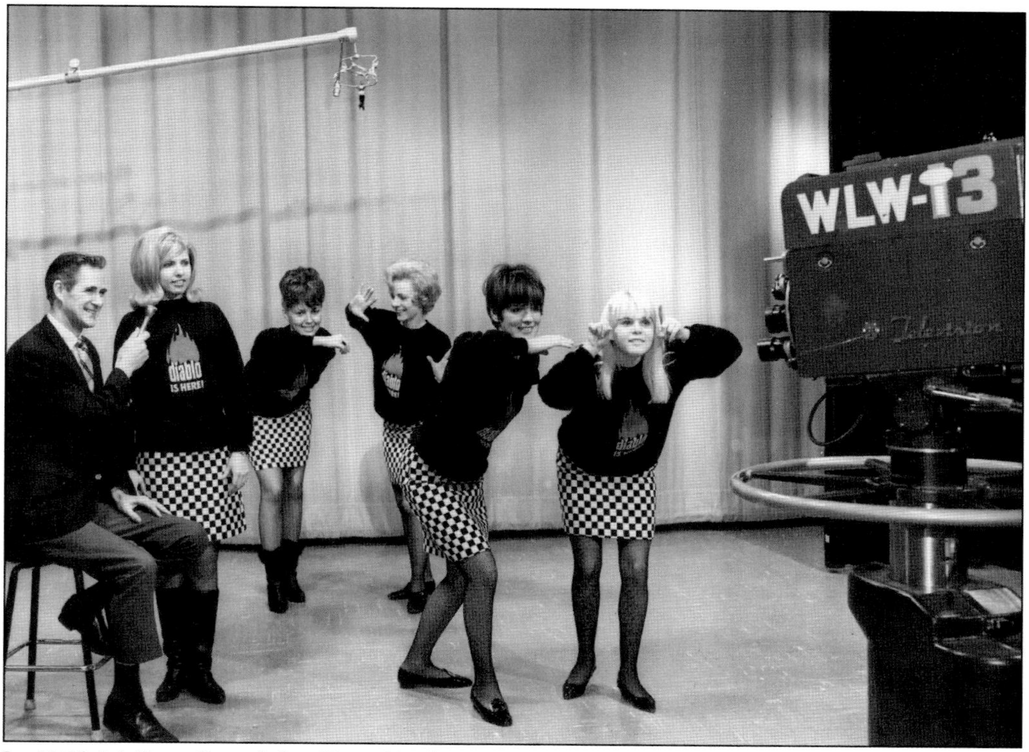

In 1965, McDowell took his *Teen Twirl* concept to WLW-I, which at the time was the ABC television affiliate in Indianapolis. He retooled it to be a kindred spirit of Dick Clark's *American Bandstand*, calling it *Bandstand 13*. This image is from around 1968.

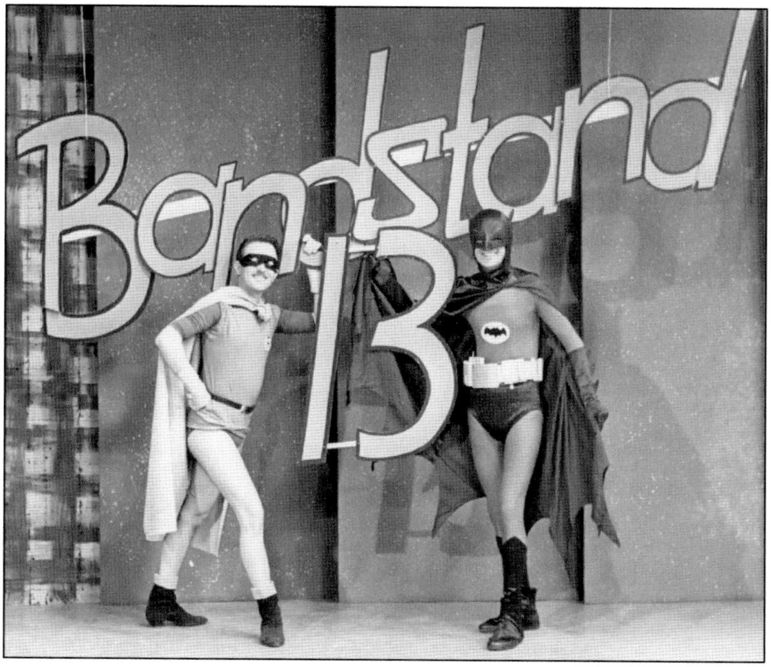

In a tribute to the mid-1960s *Batman* television series, McDowell dons the caped crusader costume, with Bill McPherson as Robin. *Batman* had been the surprise hit of the year, with "Batmania" sweeping the country.

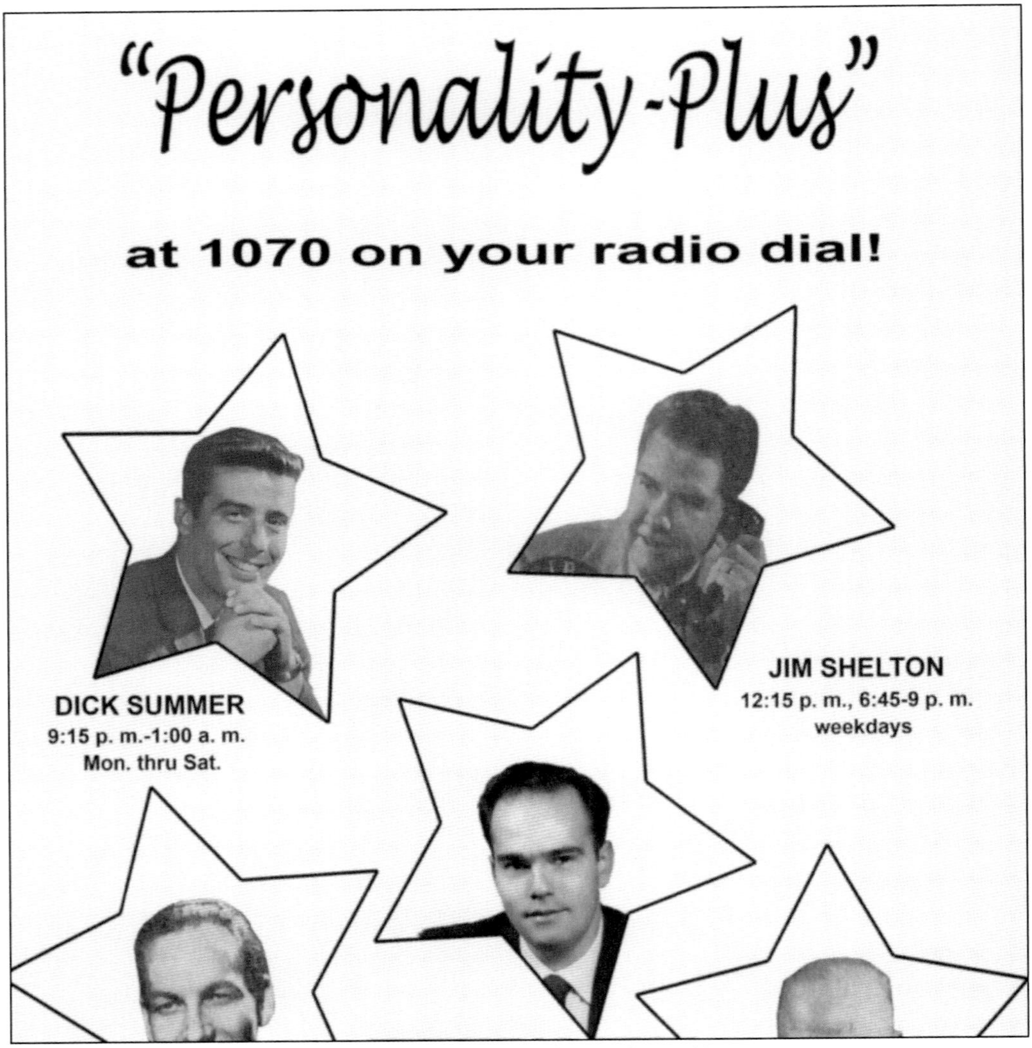

At WIBC, the emphasis on personalities continued into the 1960s.

At 6:45, Jim Shelton would come on with a program called The Platter Party, *because back in those days, records were referred to as platters. And then, when he went off at 9:00, Dick Summer had a show called* Summertime. —Art Vuolo Jr.

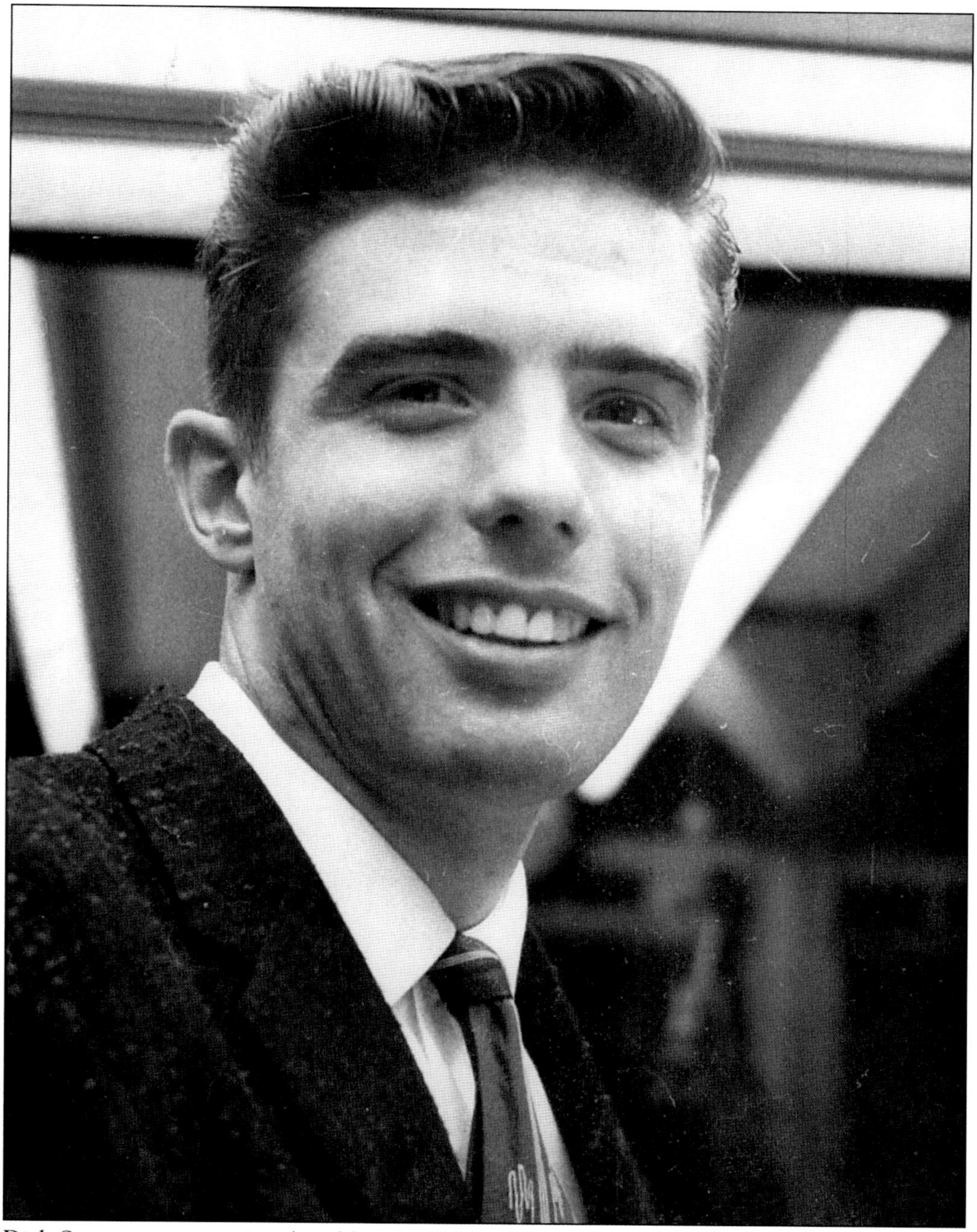

Dick Summer went national and became a broadcasting legend. He had the unique warmth, sensitivity, and style of "Wolfman Jack" and smoothness of Larry Lujack. Summer left WIBC for stints in St. Louis and Cincinnati only to return to Indy and WISH, where he hosted a teen dance show, before heading east to WBZ-AM in Boston, a station heard coast to coast.

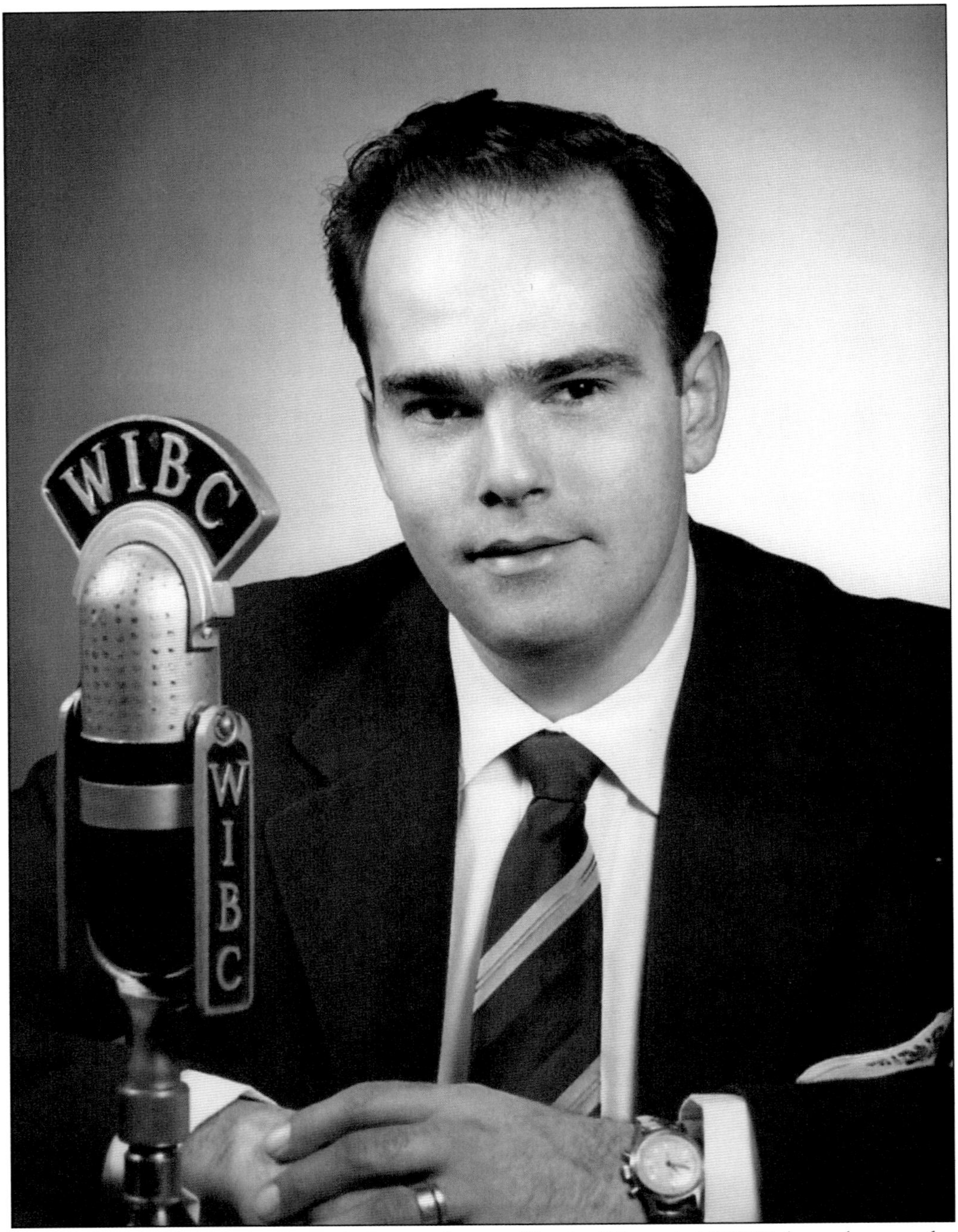

The undisputed king of radio in Indianapolis was Charles "Bouncin' Bill" Baker, who was the morning-drive disc jockey on WIBC and the city's first rock 'n' roll deejay. Bill was born in Pennsylvania in 1928 and came to WIBC via WIOU in Kokomo. At first, he worked at WIBC as an announcer doing overnight broadcasts because he had "a voice that could sell cars." He was quickly moved to the morning drive, where he pulled in very strong ratings with his mix of music and fun character voices.

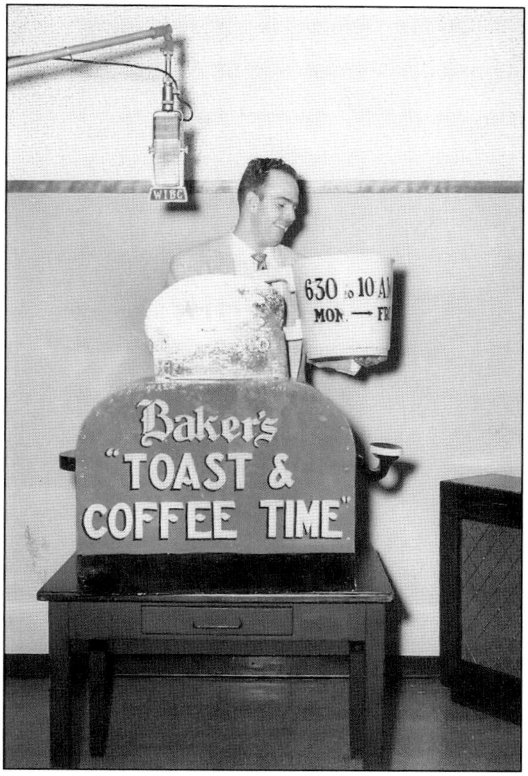

From the time Bill Baker first went on the air at WIBC, he was responsible for a significant share of the station's revenue. As early as 1956, John Blair & Company, WIBC's national marketing company, was pitching his unique show to potential sponsors.

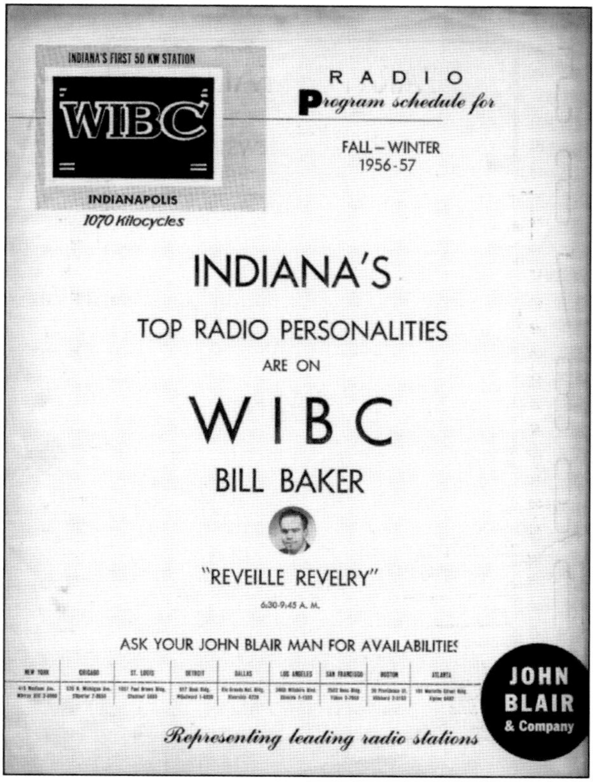

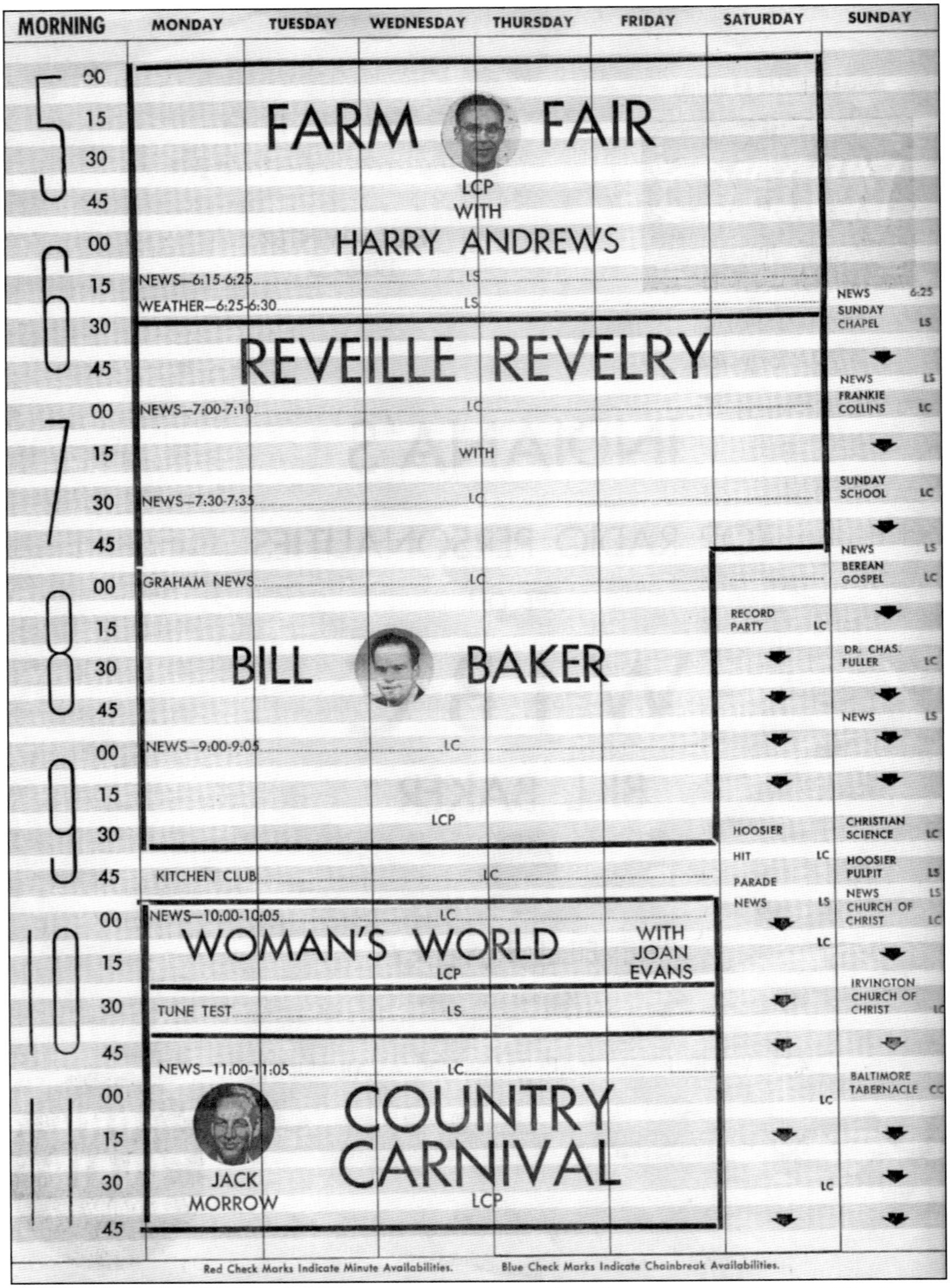
MORNING | MONDAY | TUESDAY | WEDNESDAY | THURSDAY | FRIDAY | SATURDAY | SUNDAY

5 00 15 30 45
6 00 15 30 45
7 00 15 30 45
8 00 15 30 45
9 00 15 30 45
10 00 15 30 45
11 00 15 30 45

FARM FAIR
LCP
WITH
HARRY ANDREWS
NEWS—6:15-6:25 LS
WEATHER—6:25-6:30 LS

REVEILLE REVELRY
NEWS—7:00-7:10 LC
WITH
NEWS—7:30-7:35 LC
GRAHAM NEWS LC

BILL BAKER
NEWS—9:00-9:05 LC
LCP

KITCHEN CLUB LC

NEWS—10:00-10:05 LC
WOMAN'S WORLD WITH JOAN EVANS
LCP
TUNE TEST LS

NEWS—11:00-11:05 LC
JACK MORROW
COUNTRY CARNIVAL
LCP

Saturday: RECORD PARTY LC; HOOSIER HIT PARADE LC; NEWS LS; LC; LC; LC

Sunday: NEWS 6:25; SUNDAY CHAPEL LS; NEWS LS; FRANKIE COLLINS LC; SUNDAY SCHOOL LC; NEWS LS; BEREAN GOSPEL LC; DR. CHAS. FULLER LC; NEWS LS; CHRISTIAN SCIENCE LC; HOOSIER PULPIT LS; NEWS LS; CHURCH OF CHRIST LC; IRVINGTON CHURCH OF CHRIST LC; BALTIMORE TABERNACLE CC

Red Check Marks Indicate Minute Availabilities. Blue Check Marks Indicate Chainbreak Availabilities.

The name of Bill's first morning show was *Revielle Revelry*. The nickname "Bouncin' Bill" came from the animated way he would sit on the drum stool while playing the drums.

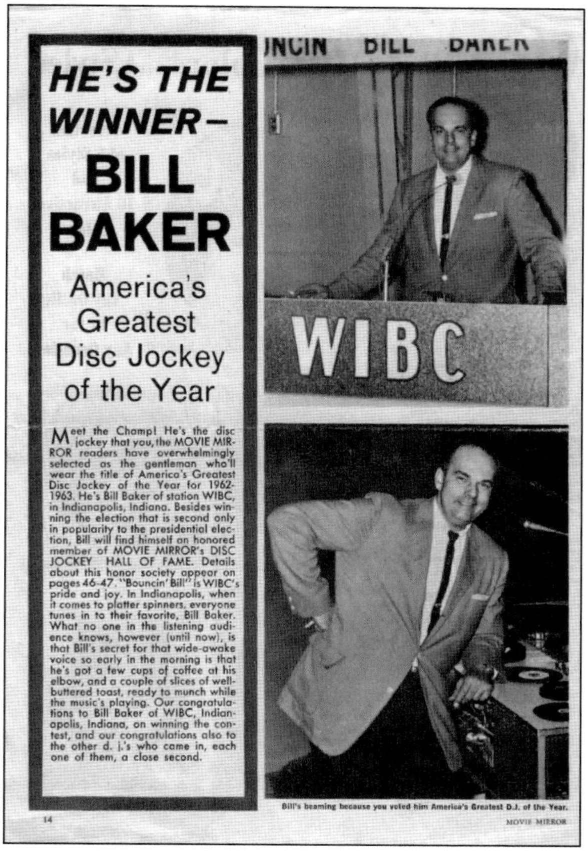

HE'S THE WINNER –

BILL BAKER

America's Greatest Disc Jockey of the Year

Meet the Champ! He's the disc jockey that you, the MOVIE MIRROR readers have overwhelmingly selected as the gentleman who'll wear the title of America's Greatest Disc Jockey of the Year for 1962-1963. He's Bill Baker of station WIBC, in Indianapolis, Indiana. Besides winning the election that is second only in popularity to the presidential election, Bill will find himself an honored member of MOVIE MIRROR's DISC JOCKEY HALL OF FAME. Details about this honor society appear on pages 46-47. "Bouncin' Bill" is WIBC's pride and joy. In Indianapolis, when it comes to platter spinners, everyone tunes in to their favorite, Bill Baker. What no one in the listening audience knows, however (until now), is that Bill's secret for that wide-awake voice so early in the morning is that he's got a few cups of coffee at his elbow, and a couple of slices of well-buttered toast, ready to munch while the music's playing. Our congratulations to Bill Baker of WIBC, Indianapolis, Indiana, on winning the contest, and our congratulations also to the other d. j.'s who came in, each one of them, a close second.

Bill's beaming because you voted him America's Greatest D.J. of the Year.

14 MOVIE MIRROR

In 1963, Bouncin' Bill was voted "America's Greatest Disk Jockey" by *Movie Mirror* magazine. It was a prestigious award that Bill was quick to capitalize on, but more importantly it was an achievement he was proud of for the rest of his career.

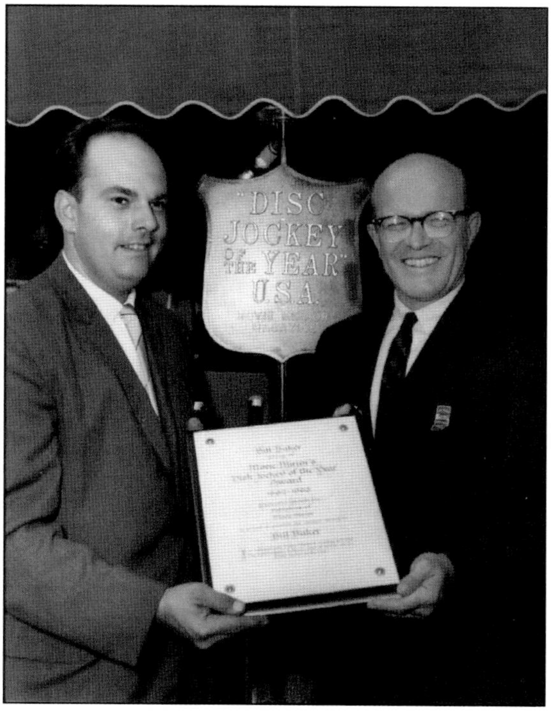

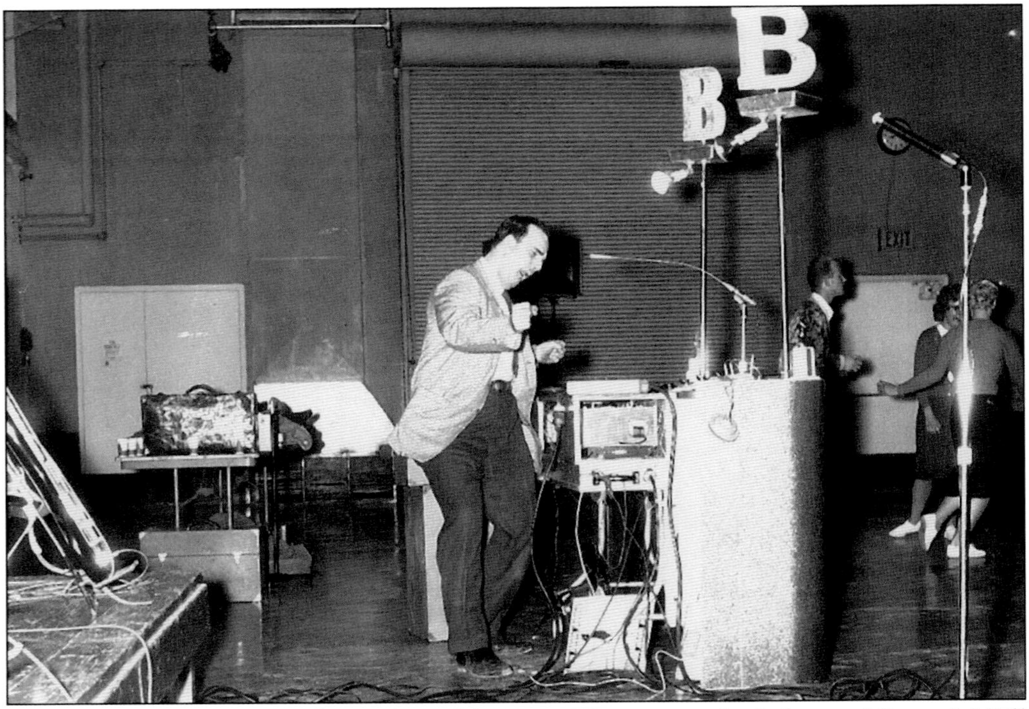

Bill Baker was the primary host of *Westlake Hops*. The event, which featured Bouncin' Bill spinning records and performances by live bands, brought in hundreds of teens every week. Paid personal appearances were a significant source of income for a radio disc jockey. For example, WNAP personality "Fast Freddie Fever" was once chastised by the comptroller at Fairbanks Broadcasting for not cashing his paychecks. He made so much making personal appearances, he didn't need the radio salary.

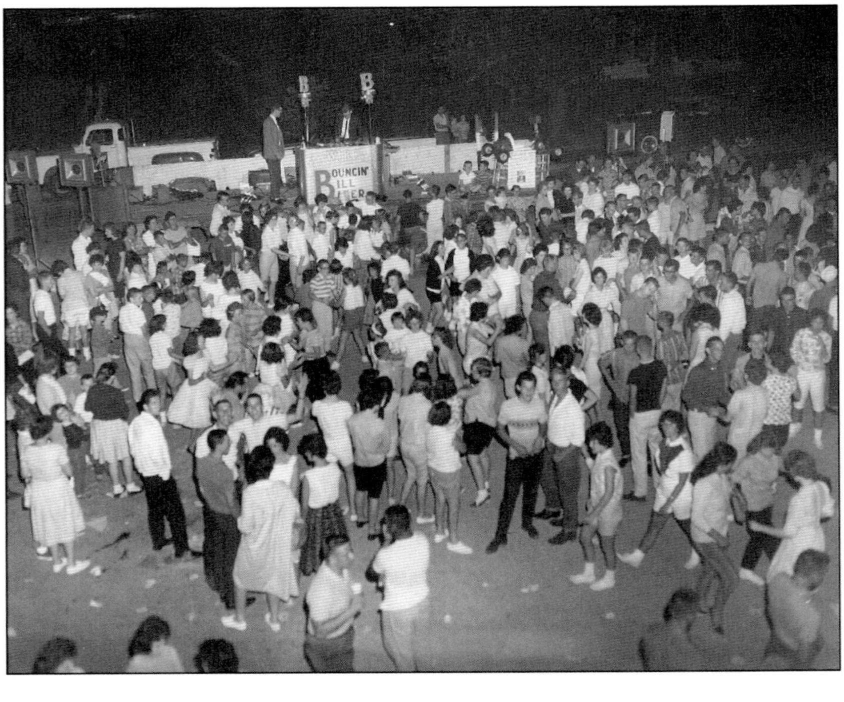

Bill Baker's show had mass appeal, particularly among teens.

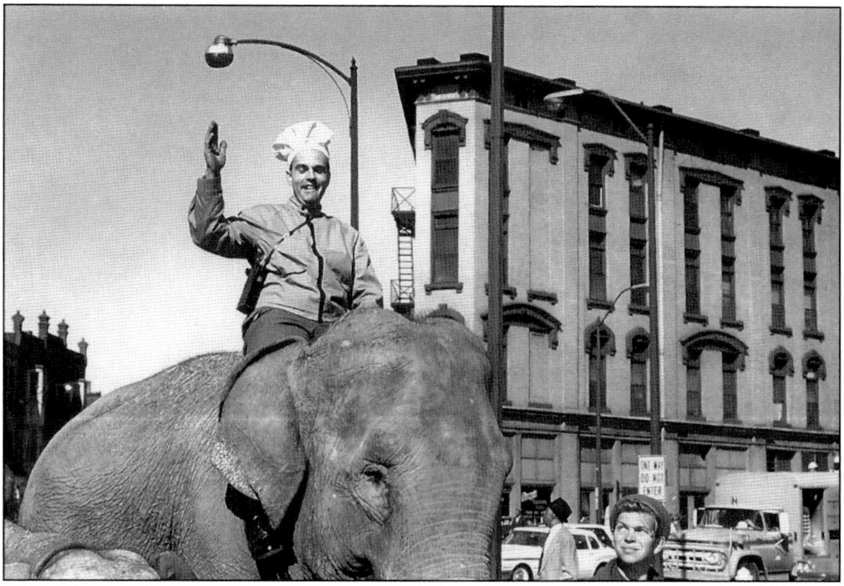

There you are, and good, good morning music lovers! You look absolutely this morning. Yours truly, the big B. Bouncin' Bill B, here on WIBC, jumping across the kilocycles, mutilating the megacycles. Here's Diane Renay, our chick of the day, and number eleven, "Navy Blue." —"Bouncin' Bill" Baker

I think Bouncin' Bill Baker was the rock 'n' roll disc jockey of that era. He was very immensely popular with his show, because he was playing very early rock 'n' roll. And that's what Bouncin' Bill Baker did. —Dave Heck (Mike "The Animal" O'Brien)

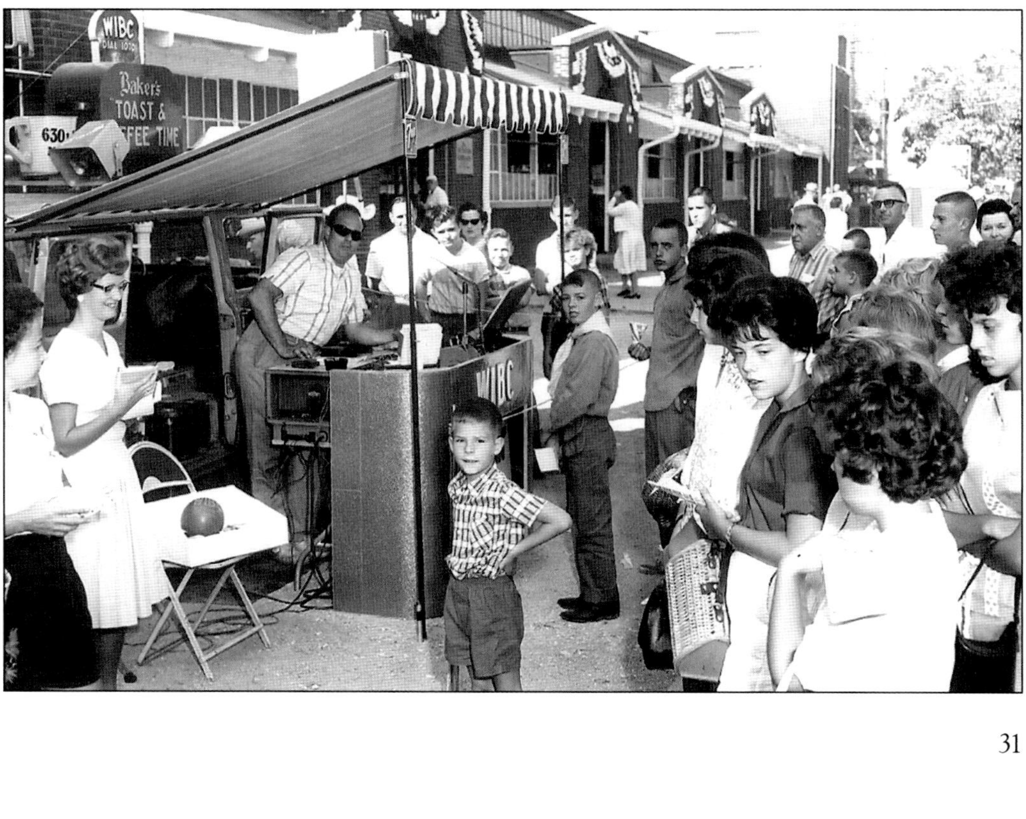

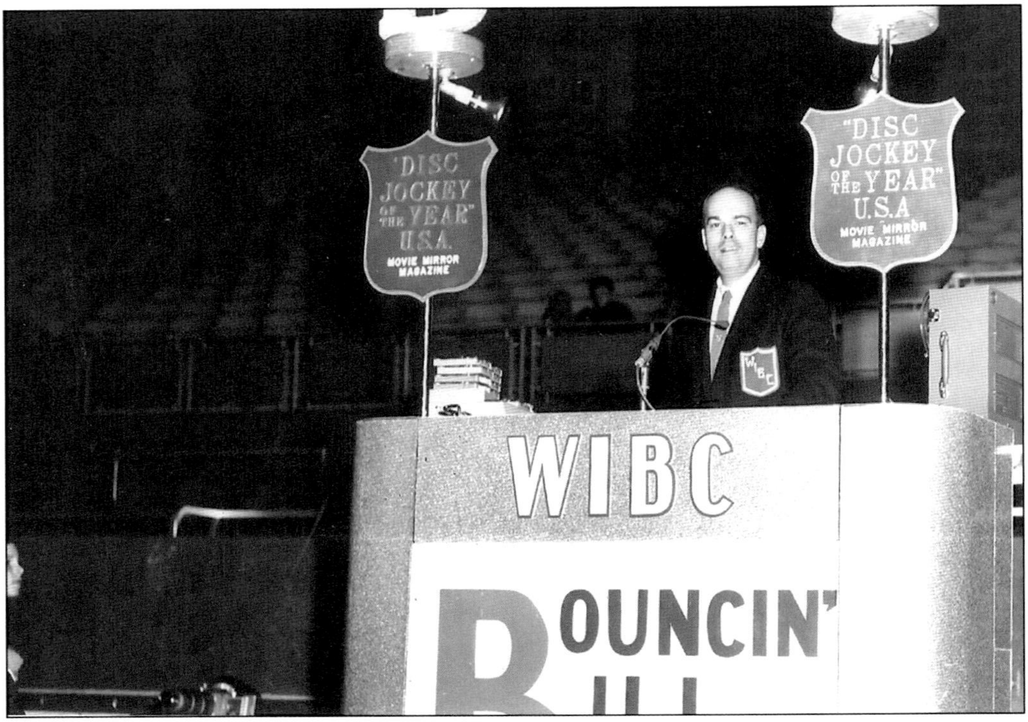

In 1969, Bill Baker retired from WIBC and formed an audio and video production company, Broadcast Productions of America. He died in August of 2005.

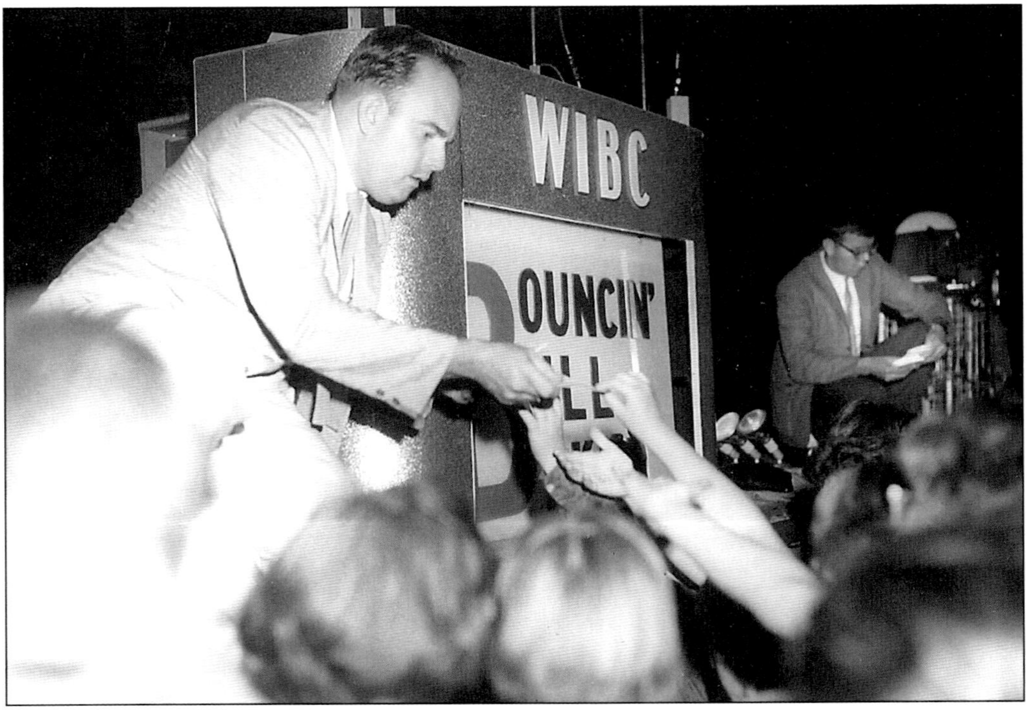

Two

The Rise of "Lucky 13" WIFE

Don Burden was a brilliant businessman and self-made millionaire. He owned KISN in Portland, Oregon, and KOIL in Omaha, Nebraska, and was looking to expand his fledgling radio empire in Indianapolis. He saw an opportunity in WISH-AM and bought the station. FM stations were considered worthless at the time, so the sellers threw WISH-FM into the deal.

Burden wanted catchy call letters and decided to call the station WIFE. WIRE thought that was too close to its own letters and filed a complaint with the FCC, asking them to decline Burden's request for WIFE. Burden instructed his attorney in Washington, DC, to work this out. At a social event in the District of Columbia, Burden's attorney happened upon an FCC commissioner and asked him about his "wire." The commissioner looked puzzled. The attorney repeated his question, "How is your wire?" Once again, the commissioner was baffled. The attorney smiled and said, "Oh, I'm sorry, I meant how is your wife? I guess there's a big difference between 'wire' and 'wife,' isn't there?" And that is how Don Burden got the FCC to approve his call letters for WIFE.

While Burden was waiting for the FCC ruling to come down, he suspended operations for WISH-AM on November 21, 1963, including its subscription to UPI news. The station temporarily changed its call letters to WAWE pending approval from the FCC for WIFE. It played music 24 hours a day, interrupted only at the top of the hour for station identification. As fate would have it, the day after dropping the subscription President Kennedy was assassinated, and WAWE had no news service. Don Burden made the decision to stick with the music-only format and forgo any news coverage. This resulted in some staff turmoil and a couple of resignations. Roughly two weeks later, WIFE hit the airwaves.

WIFE was the first true rock 'n' roll radio station in Indianapolis. —Jack Marsella

And they played the hits, and nothing but the hits, so help you that's the hits.
That's what they did. And they were the king. —Jim Hilliard

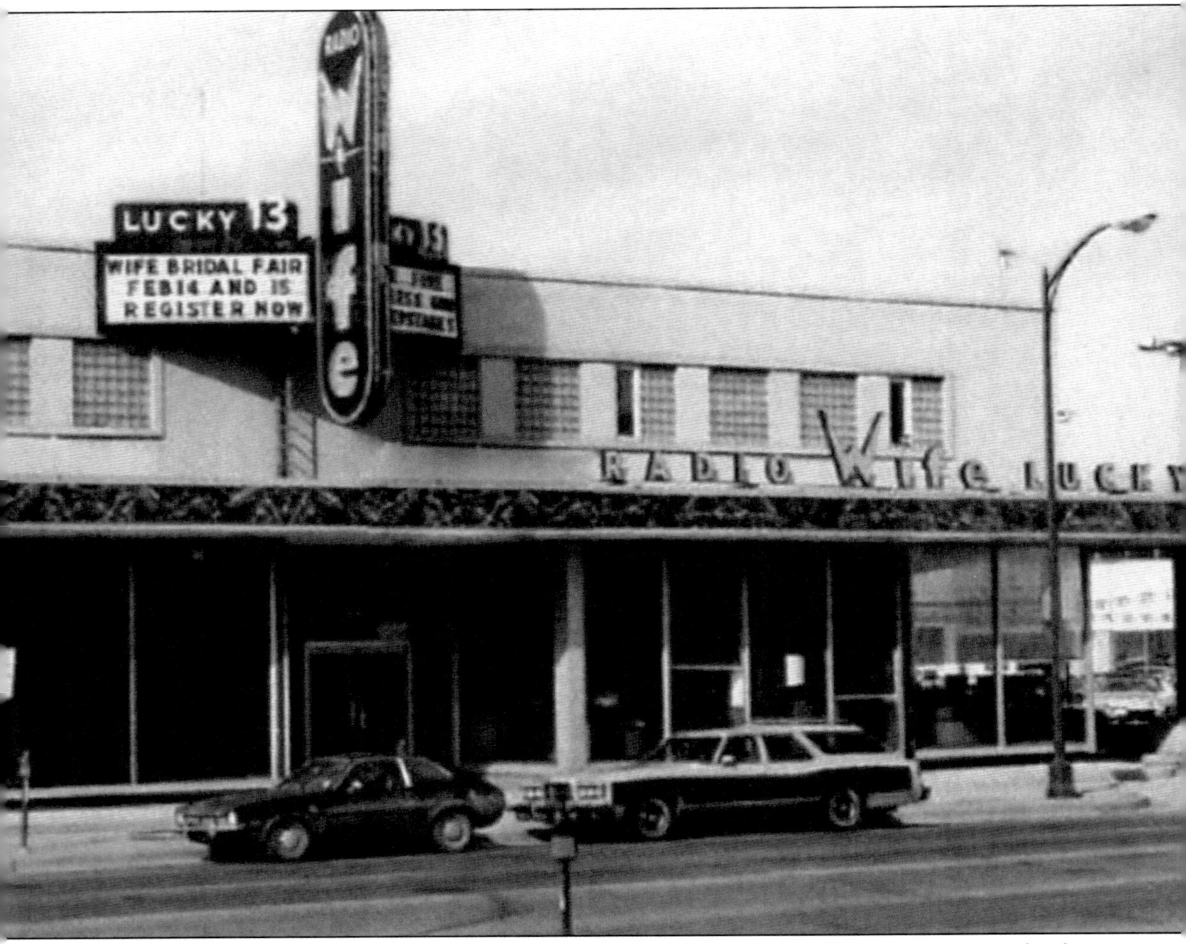

The WIFE studios were located at 1440 North Meridian Street. The most noteworthy feature was the picture window that looked out onto the sidewalk.

We had a few incidences on the "Window on the World." There were a couple of bullet holes in the window over a period of years. Nobody got hurt, fortunately. We had any number of moonings on the "Window on the World." And then, of course, Tom Mathis, the all-night guy, became very close friends with the police department in exchange for donuts. —Reb Porter

Don Burden commanded attention and respect.

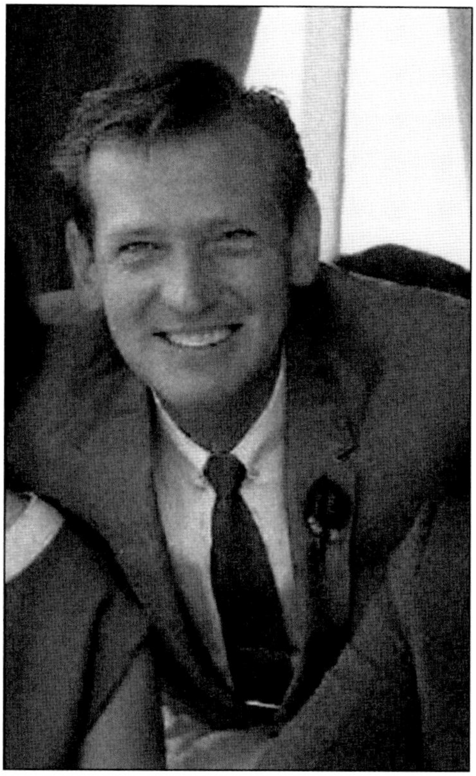

Don Burden was the kind of guy that never wanted to lose. He had an attitude like Vince Lombardi. Win. Win. Win at all times. I'm not sure the guy ever slept, 'cause he'd jump on a plane and go to a station in Omaha, Portland, so forth and so on, and then he'd be back in town for two or three days. —Pat Kelly

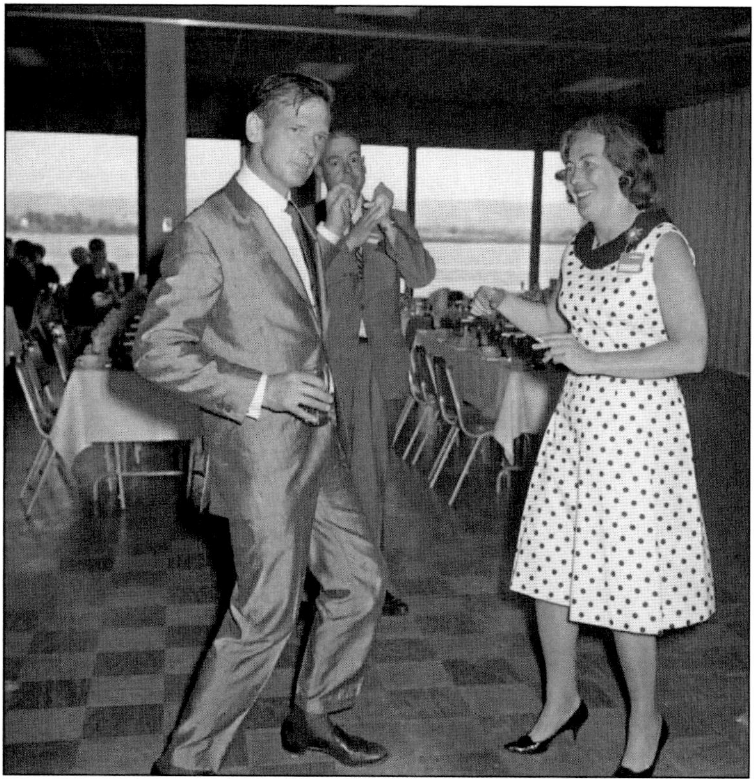

When Don Burden called me, it was never "This is Mr. B.," "This is Don Burden," "This is Mr. Burden." It was always, and in a very hushed tone, "Roger, it's me." And even when I say that today, my hands start to shake, my heart beats faster. —Roger W. Morgan

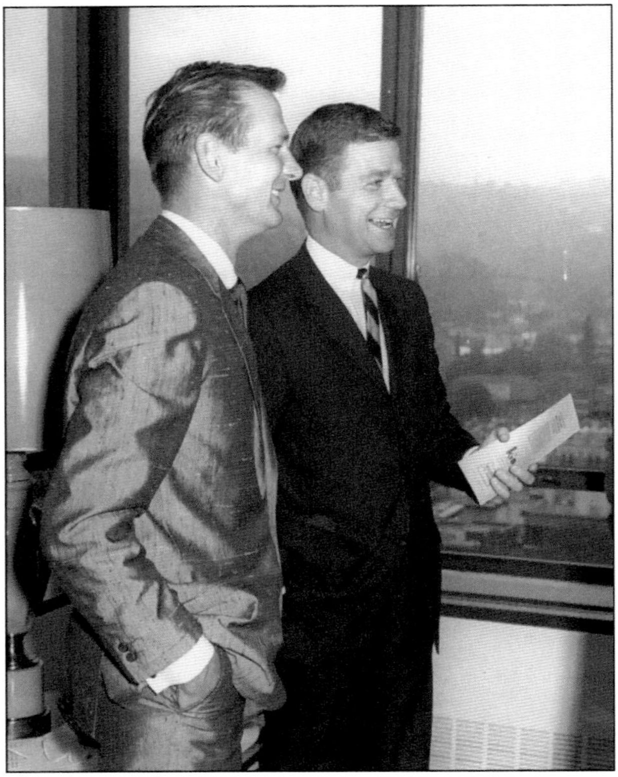

Don was a complex man of great passion.

Don Burden was a man of immediate instincts. If he was angry, he was angry as he possibly could be instantly. If he was treating you good, if he was treating you well, that was instantaneous. I could walk down the hallway, if I'd done something good for Don that day, he'd stick his hand out and shake my hand, when I pulled my hand away, there'd be a $100 bill in there. —Roger W. Morgan

Don Burden was not a person that the air staff saw very often. But when you did see him, it was memorable. One time he came into the studio, and the FM operators were supposed to be answering the request lines for the AM station because they didn't have a lot to do, they just changed the reels of tape. One time he saw they weren't being answered, he ripped the phone out of the wall, smashed it against the wall, and said, "Next time, answer the phones!" Well, of course, they couldn't answer the phones because the phone was ripped out of the wall. —Scott Wheeler

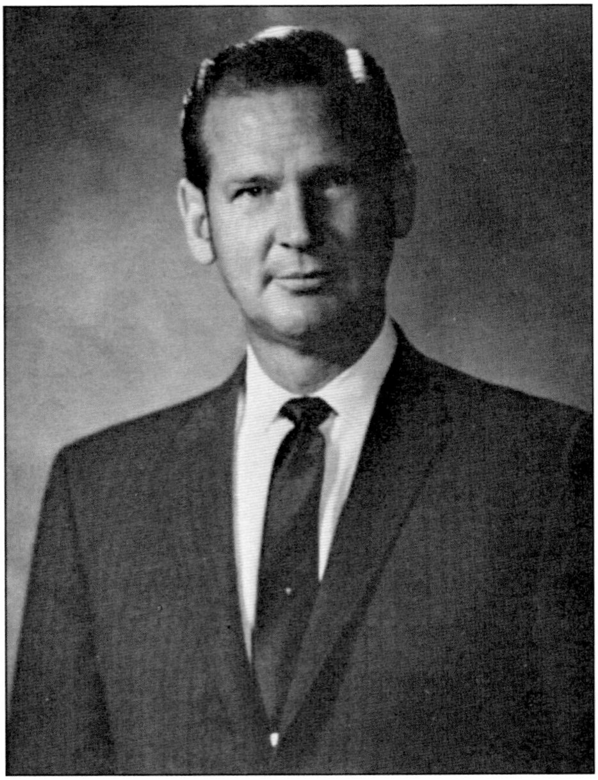

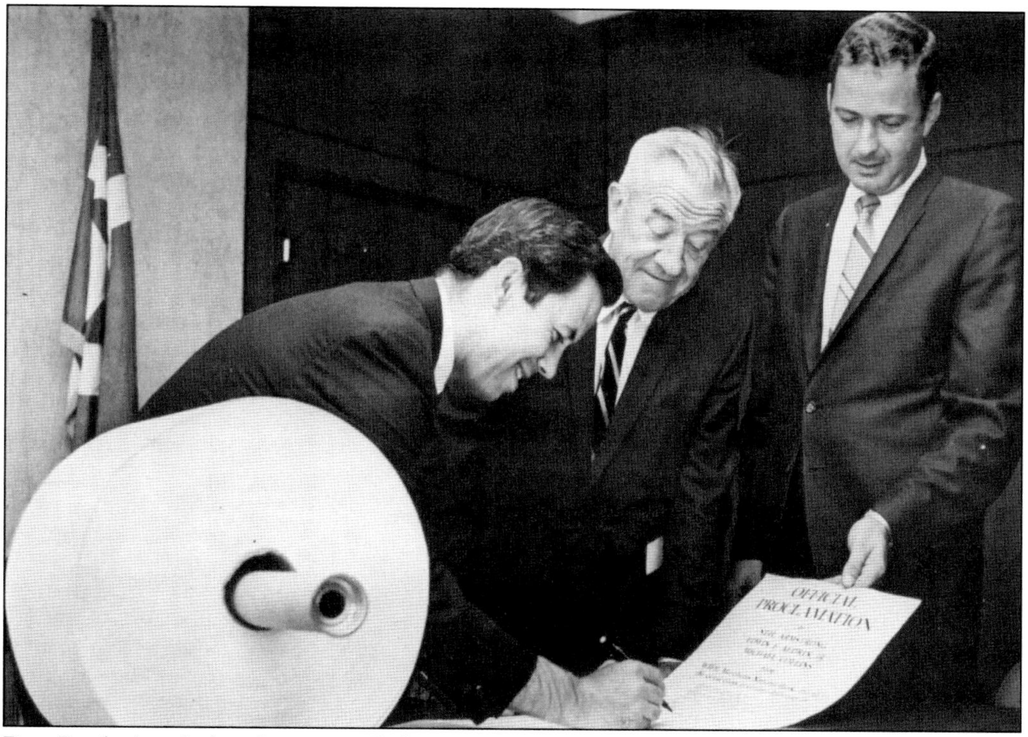

Don Burden's right-hand man was Bob Kiley (far right). As vice president and general manager at WIFE, Kiley was considered a likable and reasonable businessman. He represented the station in official business and sat on several boards throughout the city. He is seen here with Mayor Richard Lugar (left) and Merchants Bank president R.E. Sweeney Jr. at a signing of an official proclamation honoring the crew of Apollo 11.

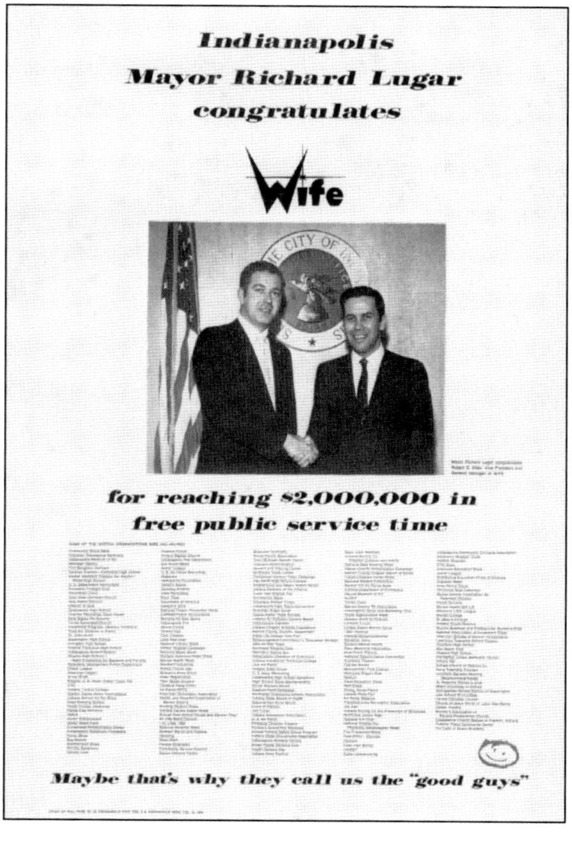

Bob Kiley (left) is pictured with Indianapolis mayor Richard Lugar being congratulated for having reached the $2 million milestone in providing free public-service time.

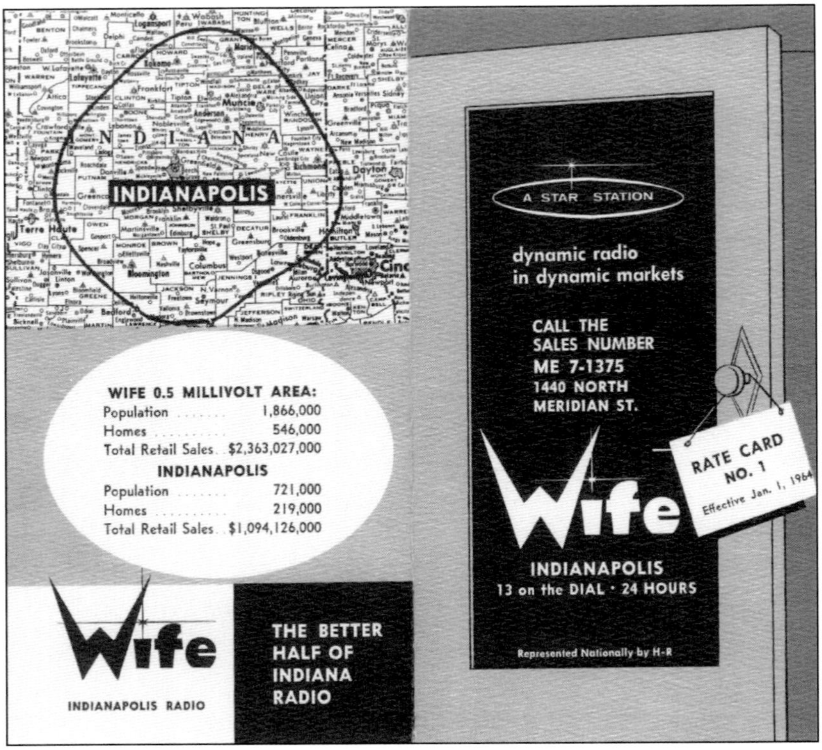

These images show "Rate Card no. 1" and the first rates for advertisement buys for January 1, 1964.

WIFE RATE CARD

CLASS AA

6:00 a.m. - 9:00 a.m. — 3:30 p.m. - 6:30 p.m.

	500	312	260	156	104	52	1
1-Min. Announcement	$17.00	$18.00	$19.00	$20.00	$21.00	$22.00	$23.00
Headline News on Half Hour	20.00	21.00	22.00	23.00	24.00	25.00	26.00
5-Min. News	26.00	27.00	28.00	29.00	30.00	31.00	32.00
10-Min. News	40.00	41.00	42.00	43.00	44.00	45.00	46.00

CLASS A

5:00 a.m. - 6:00 a.m. — 10:00 a.m. - 4:00 p.m. — 7:00 p.m. - 9:00 p.m. and All Day Sunday

	500	312	260	156	104	52	1
1-Min. Announcement	$15.00	$16.00	$17.00	$18.00	$19.00	$20.00	$21.00
Headline News on Half Hour	17.00	18.00	19.00	20.00	21.00	22.00	23.00
5-Min. News	21.00	22.00	23.00	24.00	25.00	26.00	27.00
10-Min. News	29.00	30.00	31.00	32.00	33.00	34.00	35.00

CLASS B

9:00 p.m. - 1:00 a.m.

	500	312	260	156	104	52	1
1-Min. Announcement	$ 8.00	$ 9.00	$10.00	$11.00	$12.00	$13.00	$14.00
Headline News on Half Hour	9.00	10.00	11.00	12.00	13.00	14.00	15.00
5-Min. News	12.00	13.00	14.00	15.00	16.00	17.00	18.00

SATURATION RATES

SHORT TERM — CLASS B

25	1 Min. @ $10.00
50	1 Min. @ 8.00
100	1 Min. @ 6.00

Above Packages Must Be Used In One Month

WEEKEND SATURATION

10 1-Min. Announcements $135.00 flat (Saturdays-Sundays)

B.T.A. LONG TERM (1 year)

2000 or more	1 Min. @ $12.00
1000	1 Min. @ 13.00
500	1 Min. @ 14.00
300	1 Min. @ 15.00

10 Sec. or less Announcements are 50% of Minute Rate when available—30/20 Sec. Announcements are 75% of Minute Rate when available.

TOTAL AUDIENCE PLAN

60 PLAN		40 PLAN		20 PLAN	
Class AA Morning	9	Class AA Morning	6	Class AA Morning	3
Class A	24	Class A	16	Class A	8
Class AA Afternoon	9	Class AA Afternoon	6	Class AA Afternoon	3
Class B	18	Class B	12	Class B	6
60 - 1 Min. Sales Messages Per Week	$840.00	40 - 1 Min. Sales Messages Per Week	$600.00	20 - 1 Min. Sales Messages Per Week	$320.00

Must Be Used in Seven Days

There were times where Don insisted we talk to a particular client. And our suggested budget for that client was very high. And the client would say, "Oh, my god, I've never spent that much in radio in two years!" You know, he said, "Well, this is the way you do it to be successful." When we were able to convince those people to do it and use the proper amount of dollars, they were successful.
—Pat Kelly

WIFE was a sensation.

What was originally WISH-AM, which probably had a three share of the market, probably leapt up to 20 share overnight. I mean, they absolutely destroyed WIBC. WIFE was so big; it was hard to believe how big it was. A tornado couldn't have had that big of an impact on a community like they had. —Jeff Smulyan

WIFE's brand of successful marketing, successful music, successful imaging, came at the expense of only one broadcaster: WIBC and Fairbanks Broadcasting. The audience shifted from 1070 to 1310. And it didn't happen overnight. It took about a week. —Don Worsham

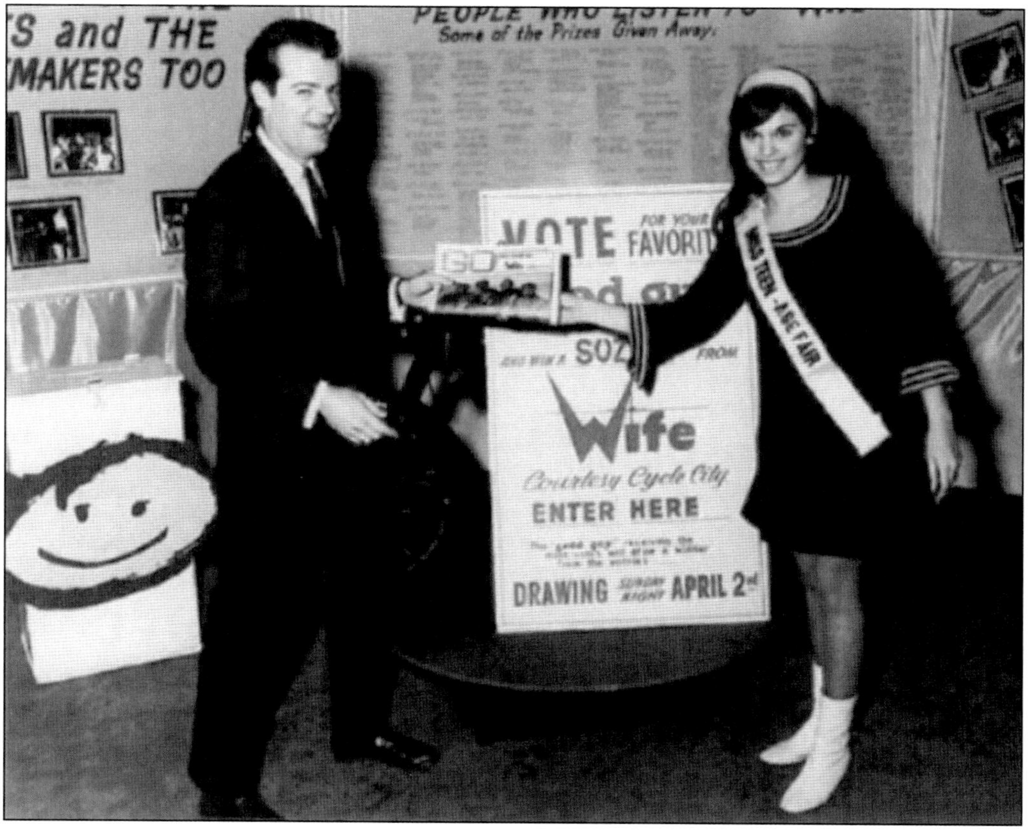

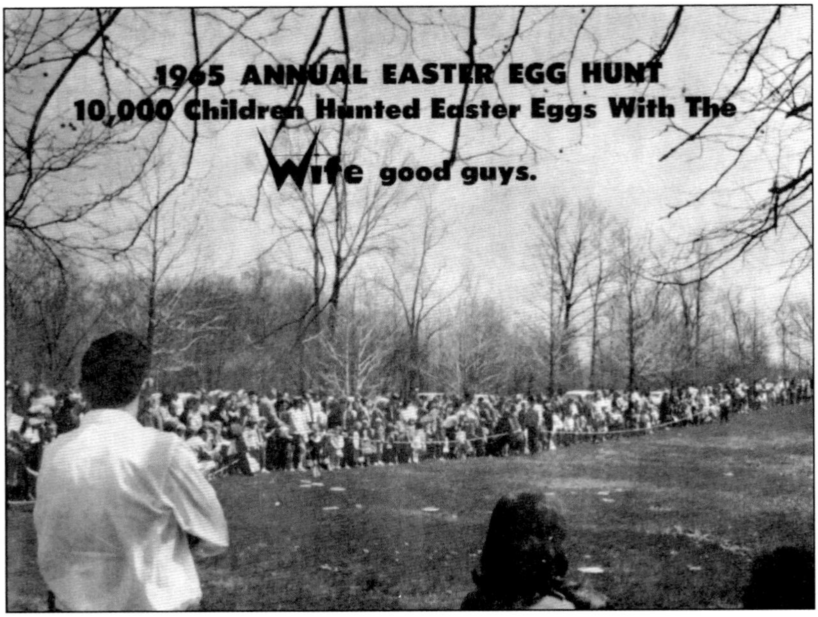

No station became better entrenched in the community than WIFE-AM.

Don Burden had a theory, and his theory was to be heavily, heavily involved in the community. To do things for people so that they knew that they had a friend in WIFE. For example, the Easter Egg Hunt. We had thousands and thousands of mothers and little kids out there, trying to pick up records, and eggs, and so forth and so on. —Pat Kelly

WIFE will furnish the following:

skateboards
25,000 Chocolate Eggs
4 Policemen (2 on the field; 2 on motorcycles
10 Transistor Radios
100 Small Stuffed Bunnies
150 Chocolate Bunnies
150 Record Albums
Hundreds of 45 RPM Records
Personnel to expedite the hunt and pass out gifts
12 Sweat shirts

Free admission tickets can be distributed at sponsor's locations.

This is another sensational WIFE promotion, of which the sponsor can take full advantage for a fraction of the cost.

Total price to sponsor will be $3,500.00

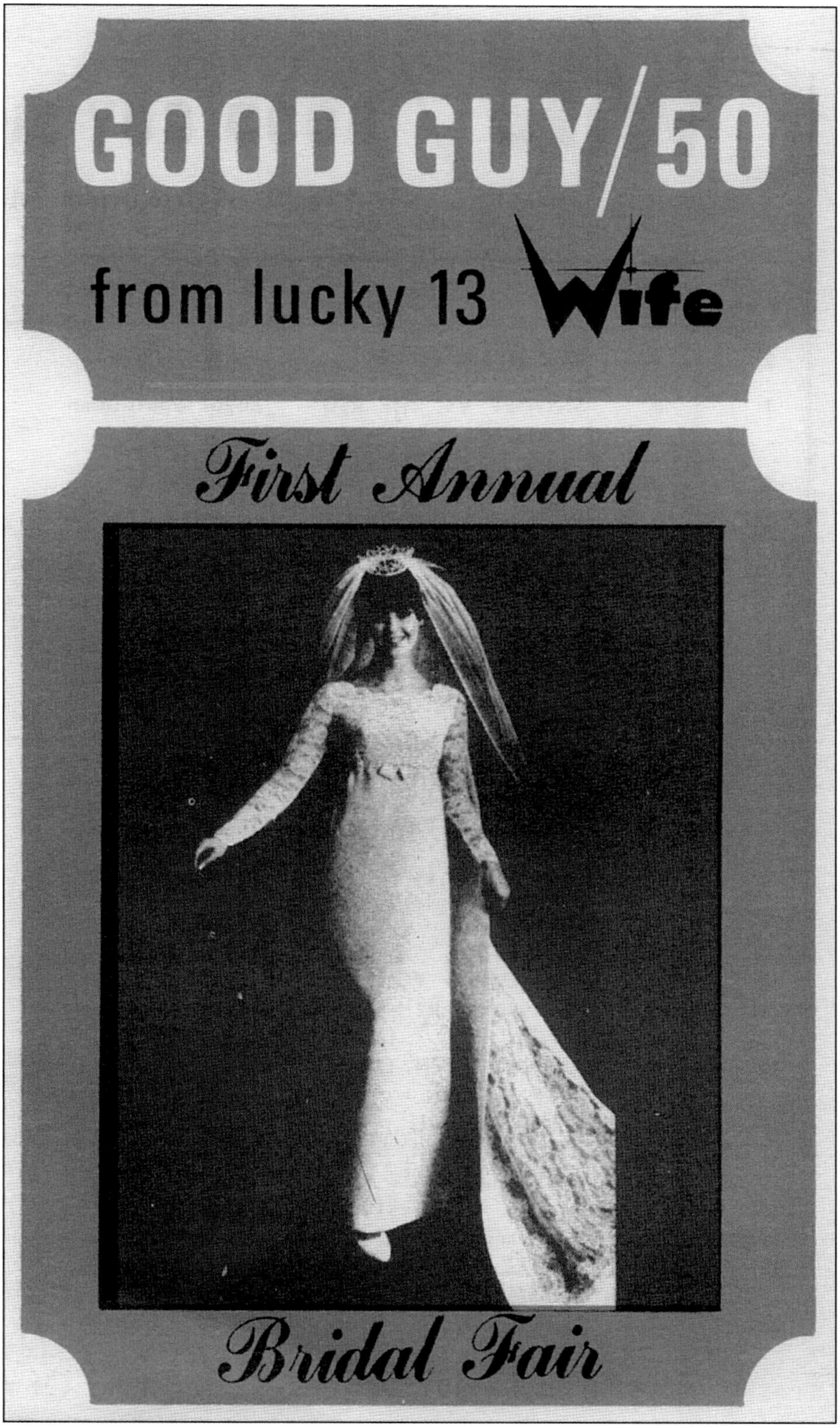

In 1967, WIFE presented its first Bridal Fair, which became a tradition for the station for several years. In addition to being able to talk to bridal experts, there was a chance to win thousands of dollars in free prizes, including clock radios, a color television, and a seven-day, six-night expense-paid honeymoon in Nassau, Bahamas.

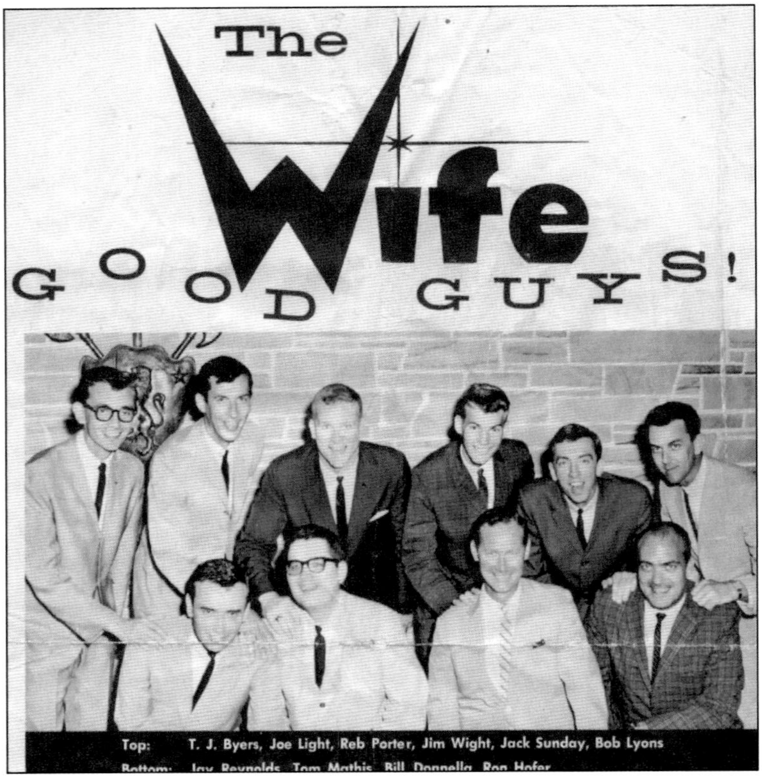

The WIFE Good Guys were superstars in Indianapolis.

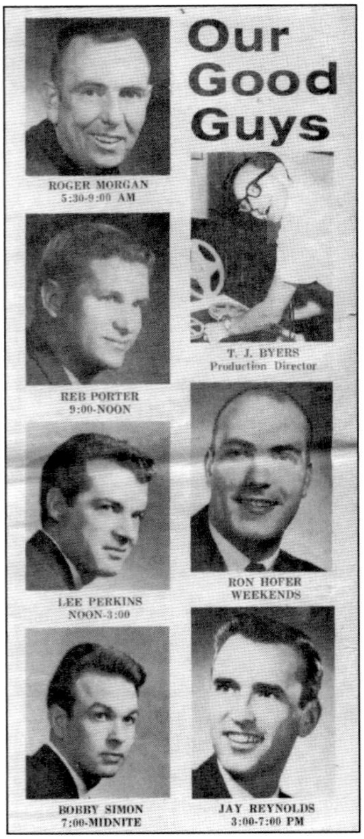

The personalities of the radio stations in the Sixties, I think, could be compared to today's rock stars. They were highly talented performers, who not only played the records, but they bantered back and forth with the audience, they had call-ins. I mean, it was quite a show. —Jack Marsella

When you're back there in a small town, the WIFE Good Guys were the equivalent of Tom Cruise. You know, it's like you're never going to see a big star. These are the biggest stars of your world. —Fast Freddie Fever

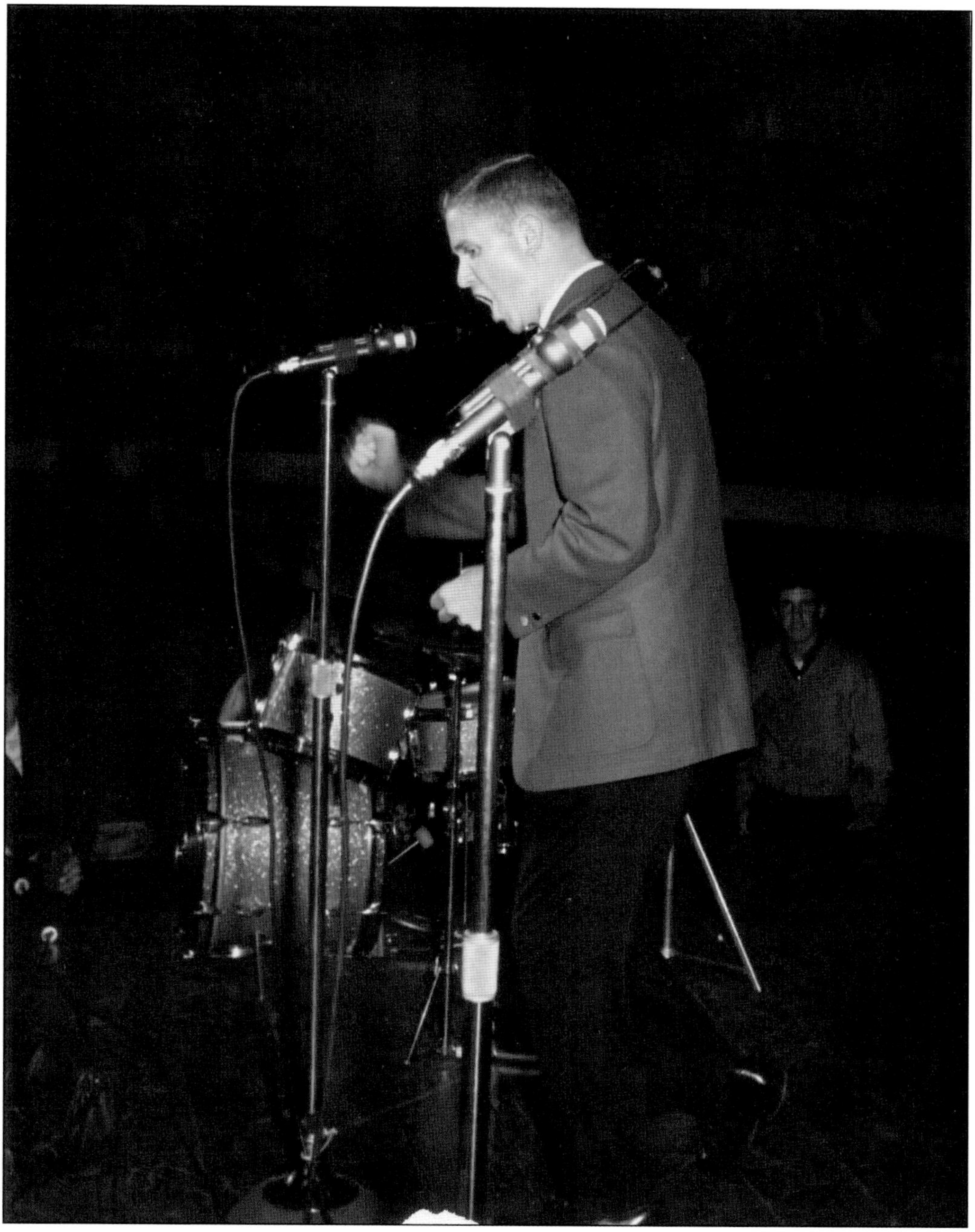

In the 1960s, the WIFE Good Guys hosted almost every concert that came through town: Tom Jones, Johnny Rivers, and Sonny & Cher, among others. Here, Reb Porter gives a rousing introduction at the Fairgrounds Coliseum.

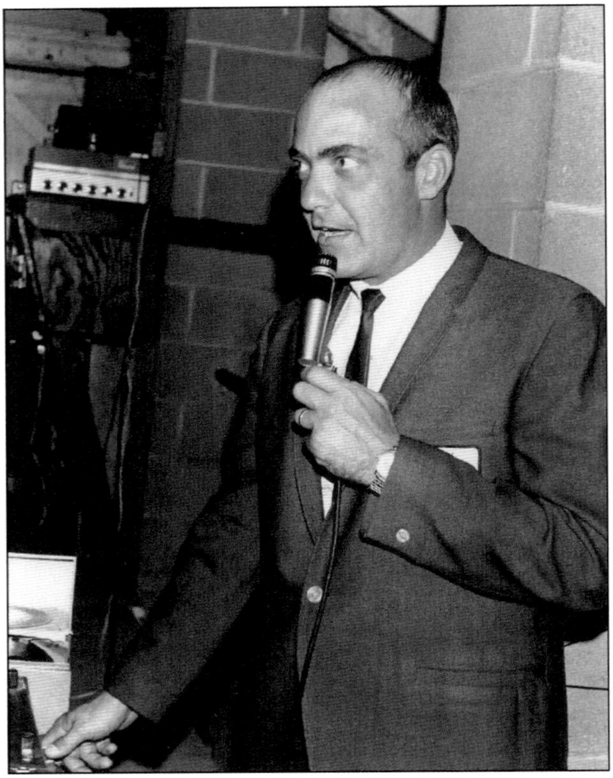

Ron Hofer held down the weekend shift and was a part-time Good Guy; during the workweek, he was a private investigator for law firms. He was once offered the program director position at the station, which he accepted. After a few months, he decided he did not like the pressure and went back to investigating.

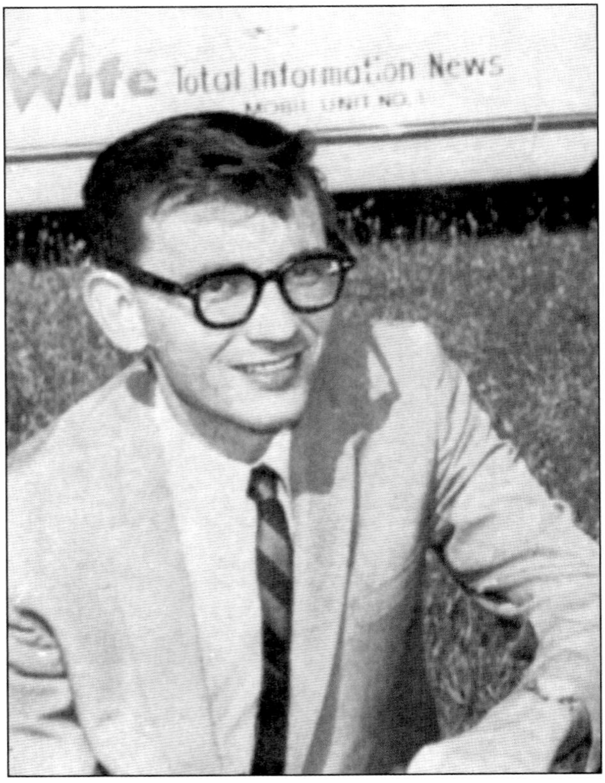

T.J. Byers had the big, deep voice on the air and was the production director. After almost 10 years, he left WIFE to go to work at WIBC.

Tom Mathis was a holdover from WISH and continued at WIFE as the overnight deejay. In 1968, he left WIFE to become general manager of the new R&B station in town, WTLC.

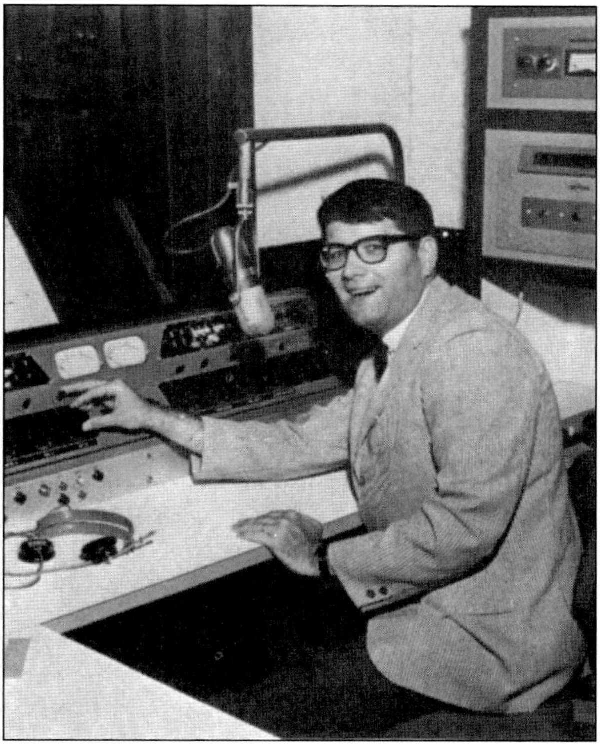

The Big Kahuna was a promotion involving a regal Polynesian king who was brought to Indianapolis. He made numerous public appearances during the summer. In reality it was Craig Martindale, a Butler University student, in costume and makeup.

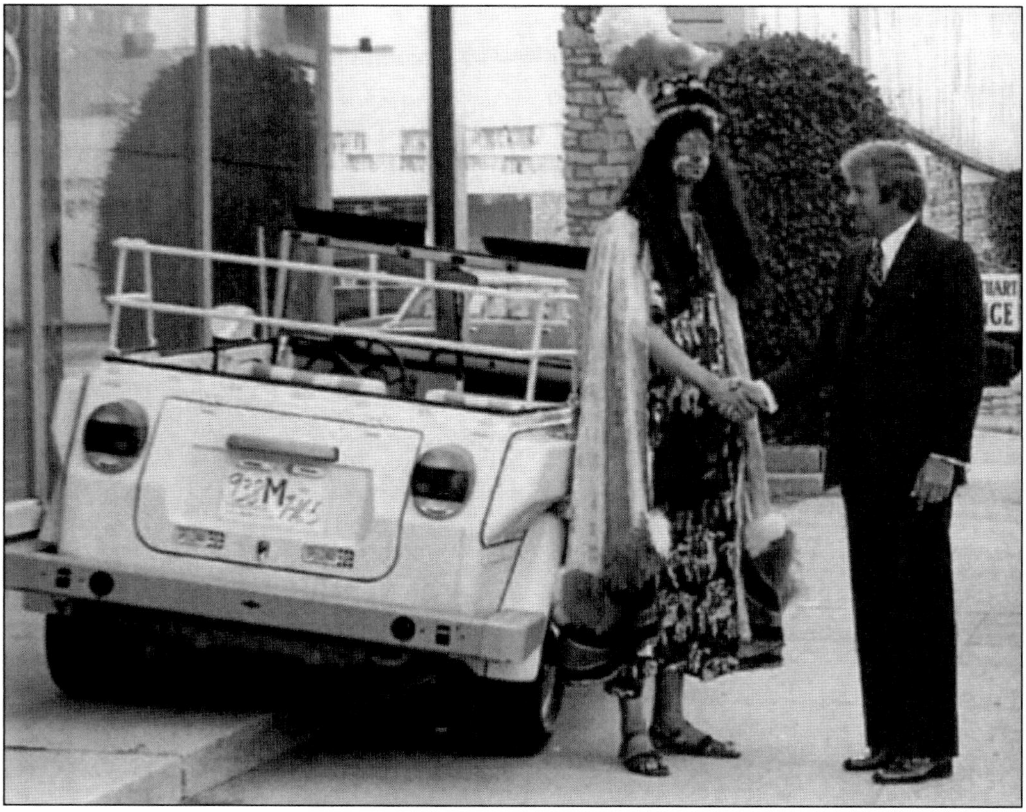

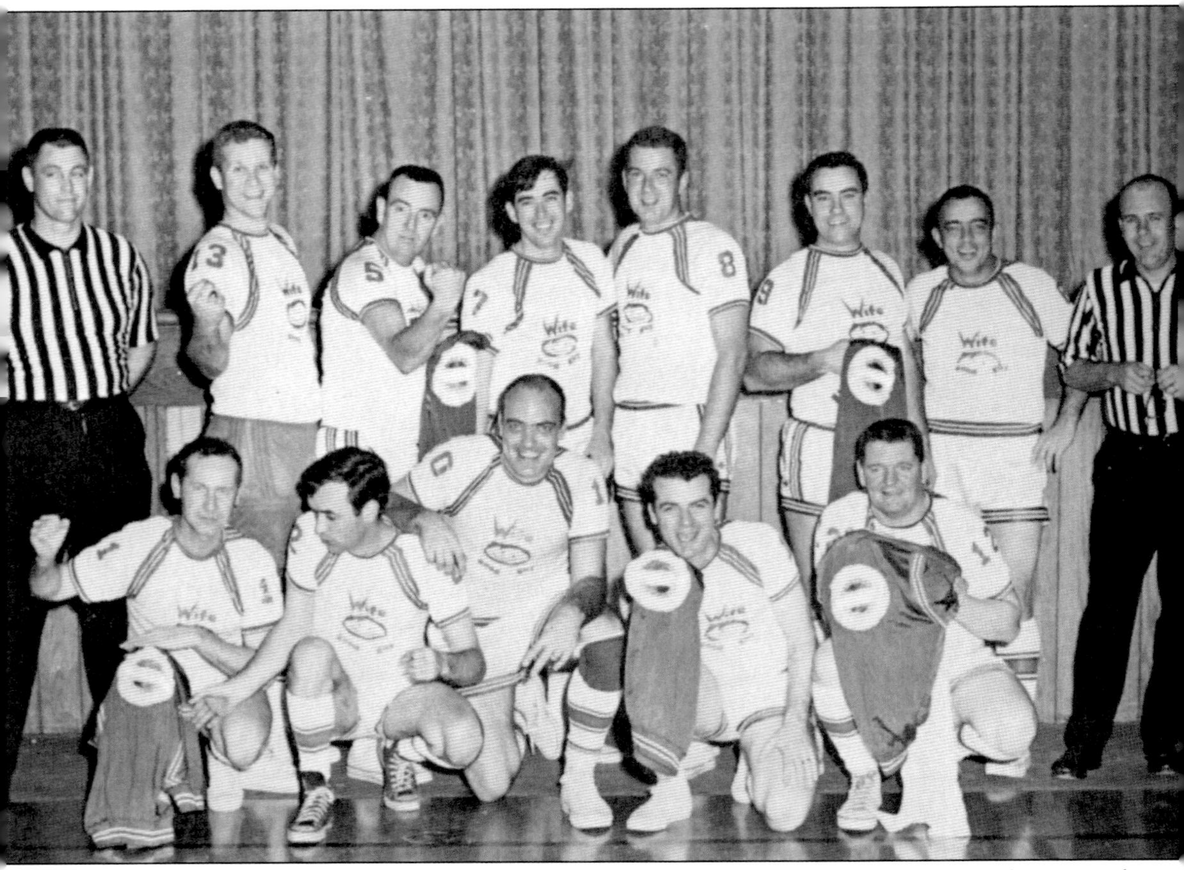

As part of its outreach to the community, WIFE assembled athletic teams made up of on-air and behind-the-scenes staff. The station raised over $40,000 for local charitable organizations through its basketball and softball teams. The teams toured central Indiana, and the station absorbed all of the expenses.

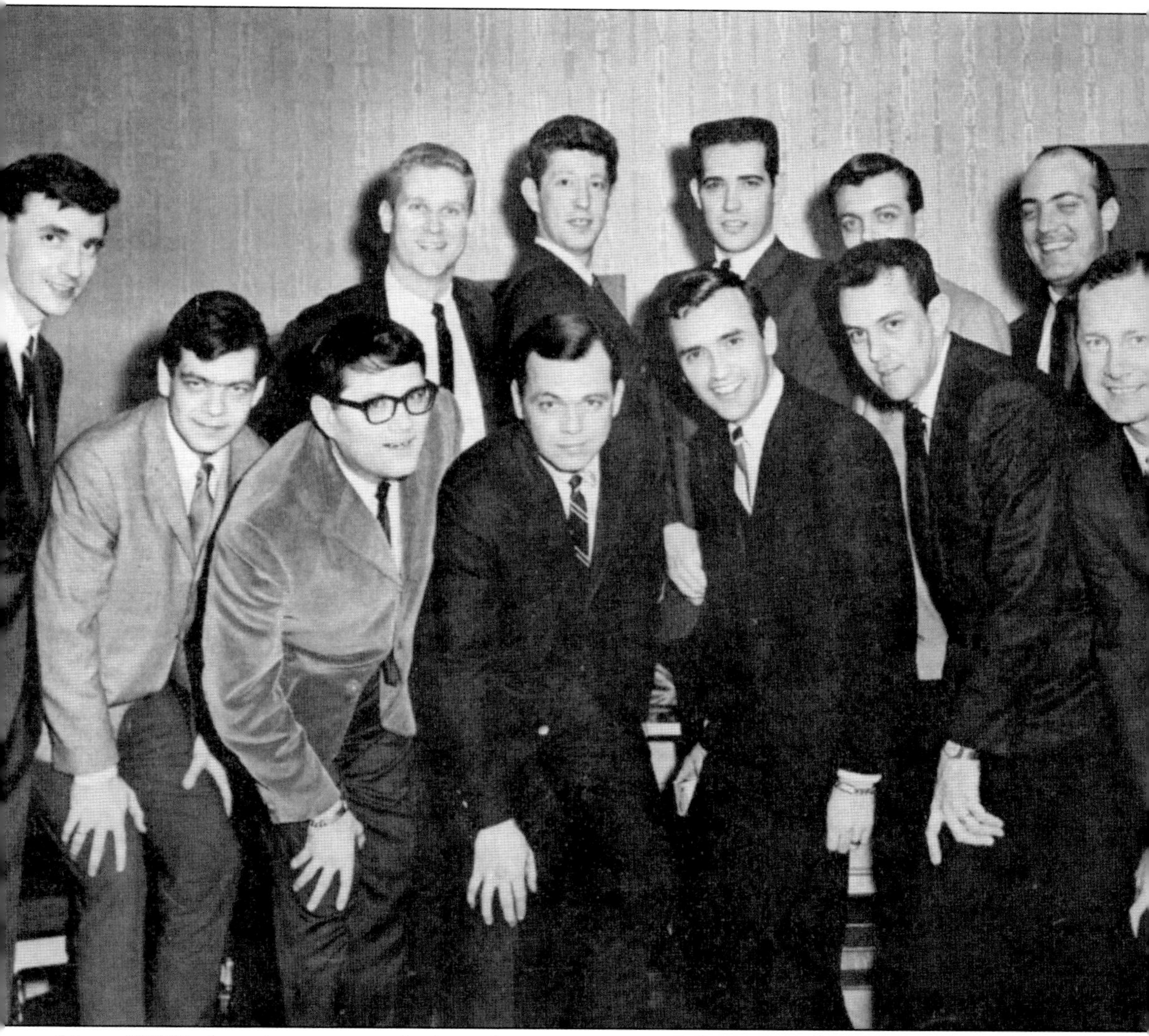

Many WIFE alumni went on to be key players in major markets. This shot of the 1966 WIFE air staff includes Fred Winston (front row, second from left), who went to WLS; Chuck "The Chucker" Browning (front row, fourth from left), to KUPD; Jay Reynolds (front row, fifth from left), to WABC; Bob Lyons (front row, second from right), to Group One; and Dick Saint (back row, second from left), to WLS.

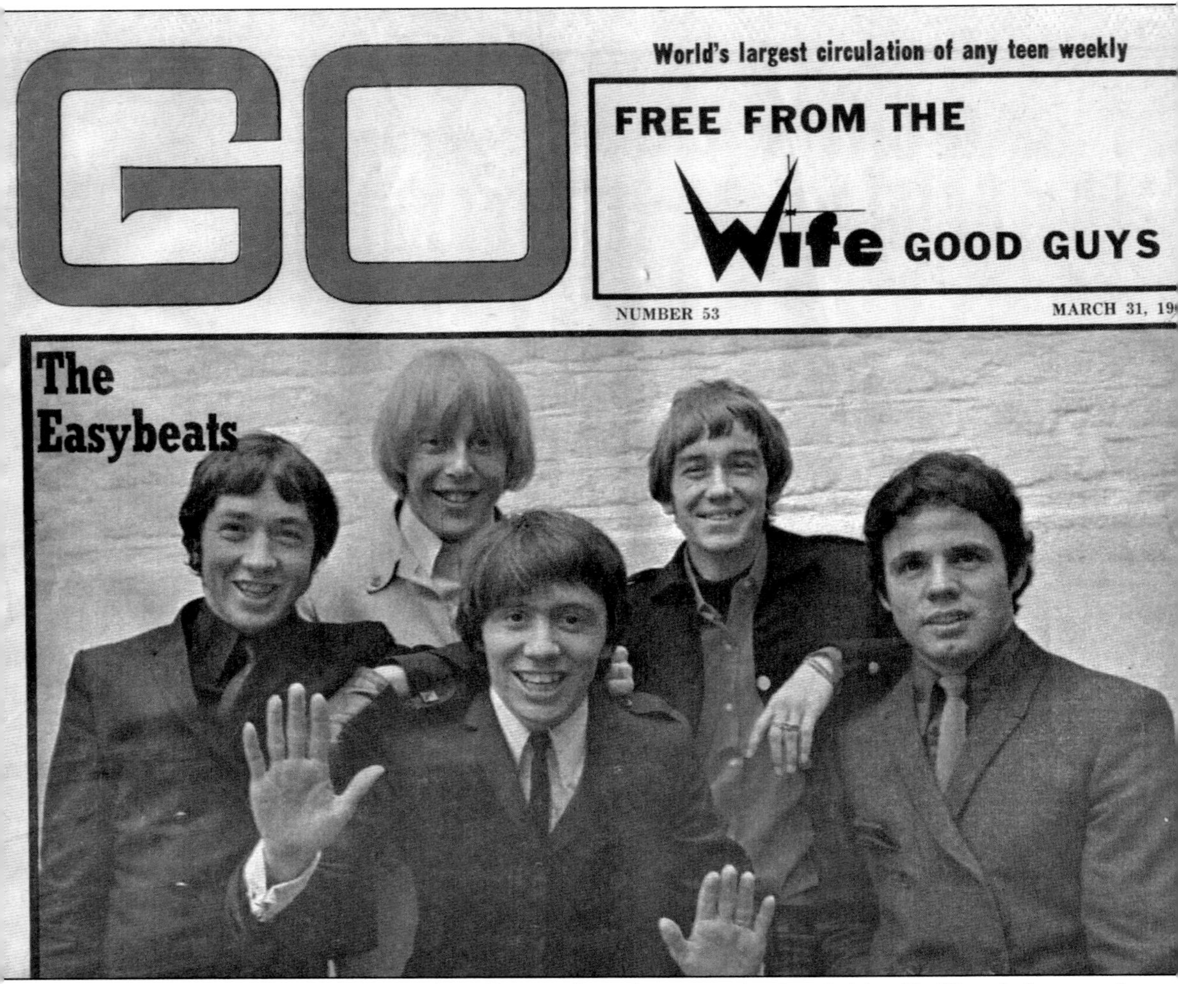

WIFE's success was partly attributable to riding the coattails of rock 'n' roll's "British Invasion" of 1964 and 1965.

I think the fact that WIFE hit the air about the time of the British Invasion probably helped them a great deal. WIFE knew how to package it and present it in a more exciting way. —Tom Cochrun

I think WIFE just rode a crest. I mean, they were just really going. They were big, and they just came in with guns a blazin'. And I think it happened so quickly that WIBC probably didn't even see that come about. I mean, WIFE brought The Beatles here. —Mike "The Animal" O'Brien

WIFE was able to put together irresistible programming.

It wasn't just the music, it's what the station was all about. And it was promotions, contests, everything. And older people, they wanted to be involved, you know. And you never knew what WIFE was going to do. —Pat Kelly

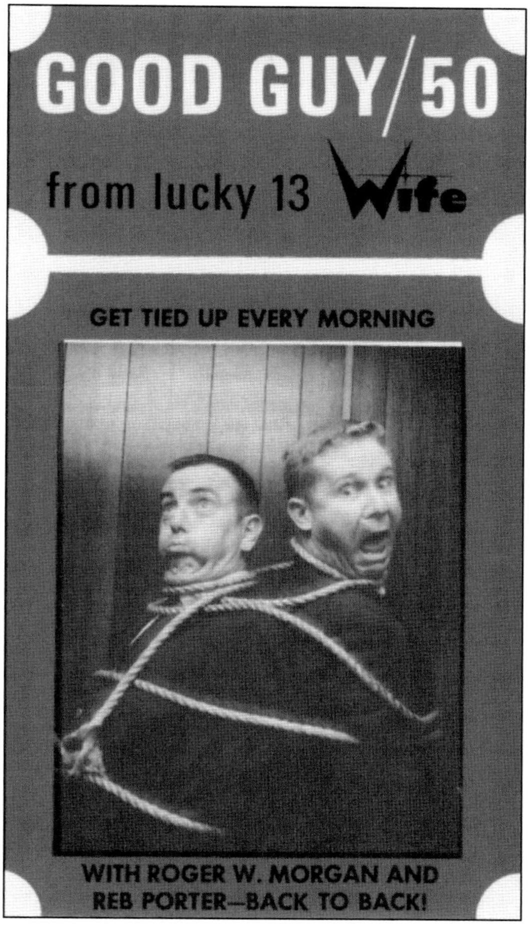

Wife / GOOD GUY SURVEY FOR THE MIDWEST

Week of October 16 thru October 23, 1968

Last Week	This Week	Title	Artist
7	1.	THOSE WERE THE DAYS	Mary Hopkins
1	2.	FIRE	Crazy A. Brown
2	3.	OVER YOU	Union Gap
6	4.	ELENORE	Turtles
5	5.	LITTLE GREEN APPLES	O. C. Smith
15	6.	WATCH TOWER	Jimi Hendrix
14	7.	HEY, WESTERN UNION MAN	Jerry Butler
3	8.	SUZIE Q	C. Clearwater
23	9.	WHITE ROOM	Cream
4	10.	MY SPECIAL ANGEL	Vogues
11	11.	PIECE OF MY HEART	Big Brother & Holding Co.
18	12.	SHAPE OF THINGS	Max Frost
25	13.	HOLD ME TIGHT	Johnny Nash
9	14.	POOR BABY	Cowsills
8	15.	REVOLUTION/HEY JUDE	Beatles
30	16.	CHAINED	Marvin Gaye
34	17.	MAGIC CARPET RIDE	Steppen Wolf
20	18.	SWEET BLINDNESS	5th Dimension
10	19.	MIDNIGHT CONFESSIONS	Grass Roots
12	20.	SLIP AWAY	Clarence Carter
24	21	RIDE SEE SAW	Moody Blues
20	22.	TAKE ME FOR A WHILE	Vanilla Fudge
29	23.	SHAKE	Shadows of Knight
36	24.	BANG SHANG A LANG	Archies
37	25.	HI-HEEL SNEAKERS/FLIP	Jose' Feliciano
DEBUT	26.	ABRAHAM, MARTIN & JOHN	Dion
13	27.	IN-A-GADDA-VIDA	Iron Butterfly
16	28.	BABY COME BACK	Equals
19	29.	I FOUND A TRUE LOVE	Wilson Pickett
22	30.	NATURALLY STONED	The Avant-Garde
21	31.	SHOOT 'EM UP, BABY	Andy Kim
DEBUT	32.	LOVE CHILD	Supremes
27	33.	GIRL WATCHER	O'Kaysions
31	34.	QUICK JOEY SMALL	Super K Circus
28	35.	I MET HER IN CHURCH	Boxtops
32	36.	FIRE	5 x 5
33	37.	I SAY A PRAYER	Aretha Franklin
39	38.	I'M IN A DIFFERENT	Four Tops
43	39.	DO THE CHOO CHOO	Archie Bell & Drells
44	40.	DO YOUR OWN THING	Brook Benton
DEBUT	41.	LALENA	Donovan
49	42.	CHEWY CHEWY	Ohio Express
48	43.	LITTLE ARROWS	Leapy Lee
47	44.	STORMY	Classics IV
46	45.	SOUL DRIPPIN'	The Mauds
41	46.	YOU GOT THE LOVE	Prof. M's Lollipop
DEBUT	47.	THE STRAIGHT LIFE	Bobby Goldsboro
DEBUT	48.	MY GROOVY BABY	Tom Dooley
DEBUT	49.	SHAME SHAME	Magic Lanterns
DEBUT	50.	PICKIN' WILD MT. BERRIES	Peggy & Jo Jo

MOST REQUESTED SONG Abraham, Martin & John — Dion

The thing about WIFE when they first hit the air that's so amazing is the signal was not all that great, but they came on very, very strong with lots of exciting personalities, lots of energy, lots of contests, lots of giveaways, and a sound that this town had not heard. For a radio station with 5,000 watts and a marginal signal to knock off 50,000-watt WIBC that had been kingpin for so long. —Art Vuolo Jr.

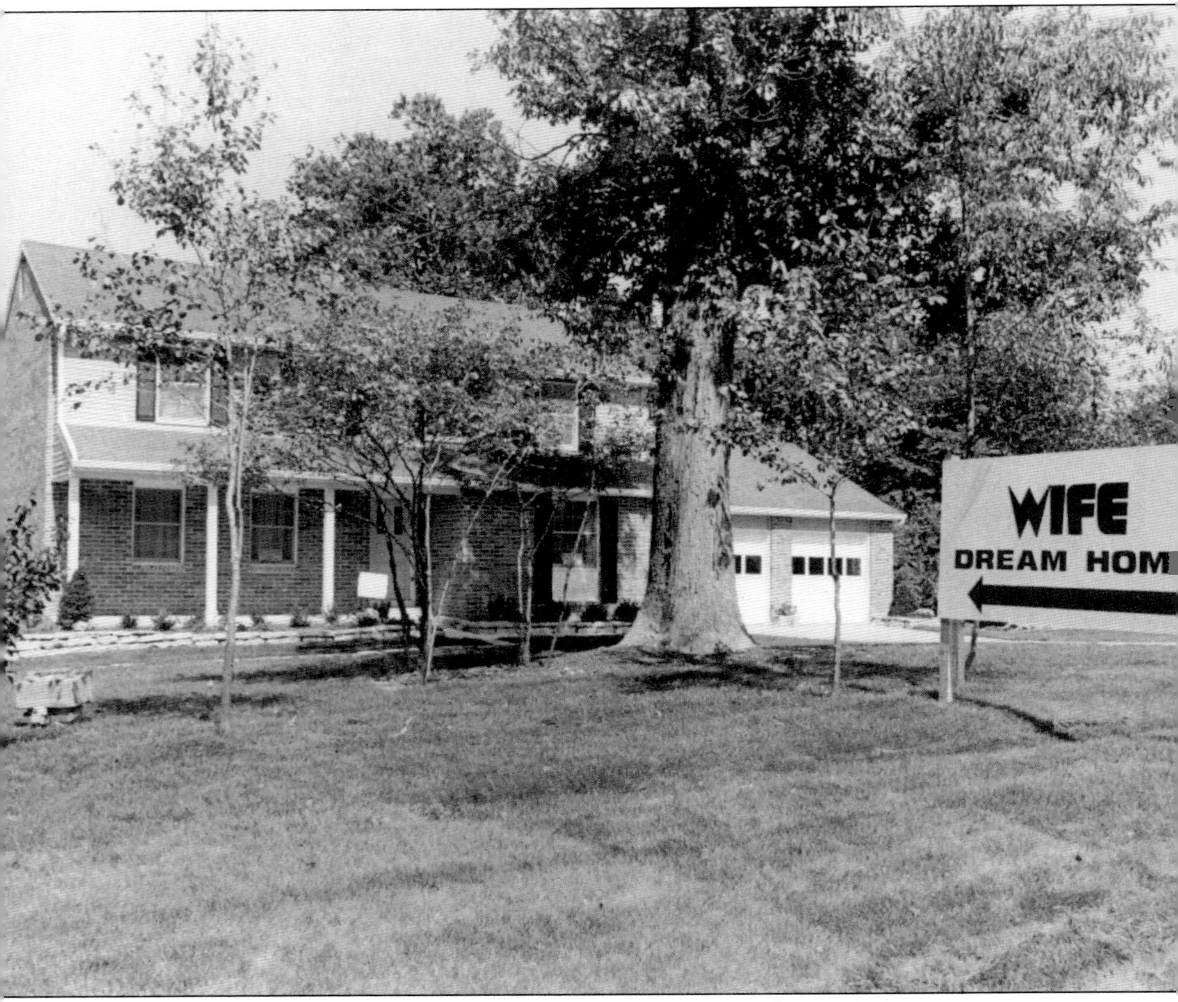

WIFE created possibly the biggest promotional giveaway in Indianapolis history.

I have to smile when I think of the Dream Home. It was built by Ryland Homes, it was totally furnished by Kittle's Furniture, and we had an on-air contest that ran for four weeks—or whatever the number of weeks it was. We drew out of a hat one winner, and did a remote live at the home where we gave it away. And one couple won a totally furnished dream home lock, stock, and barrel. That was huge. Huge. —Jack Marsella

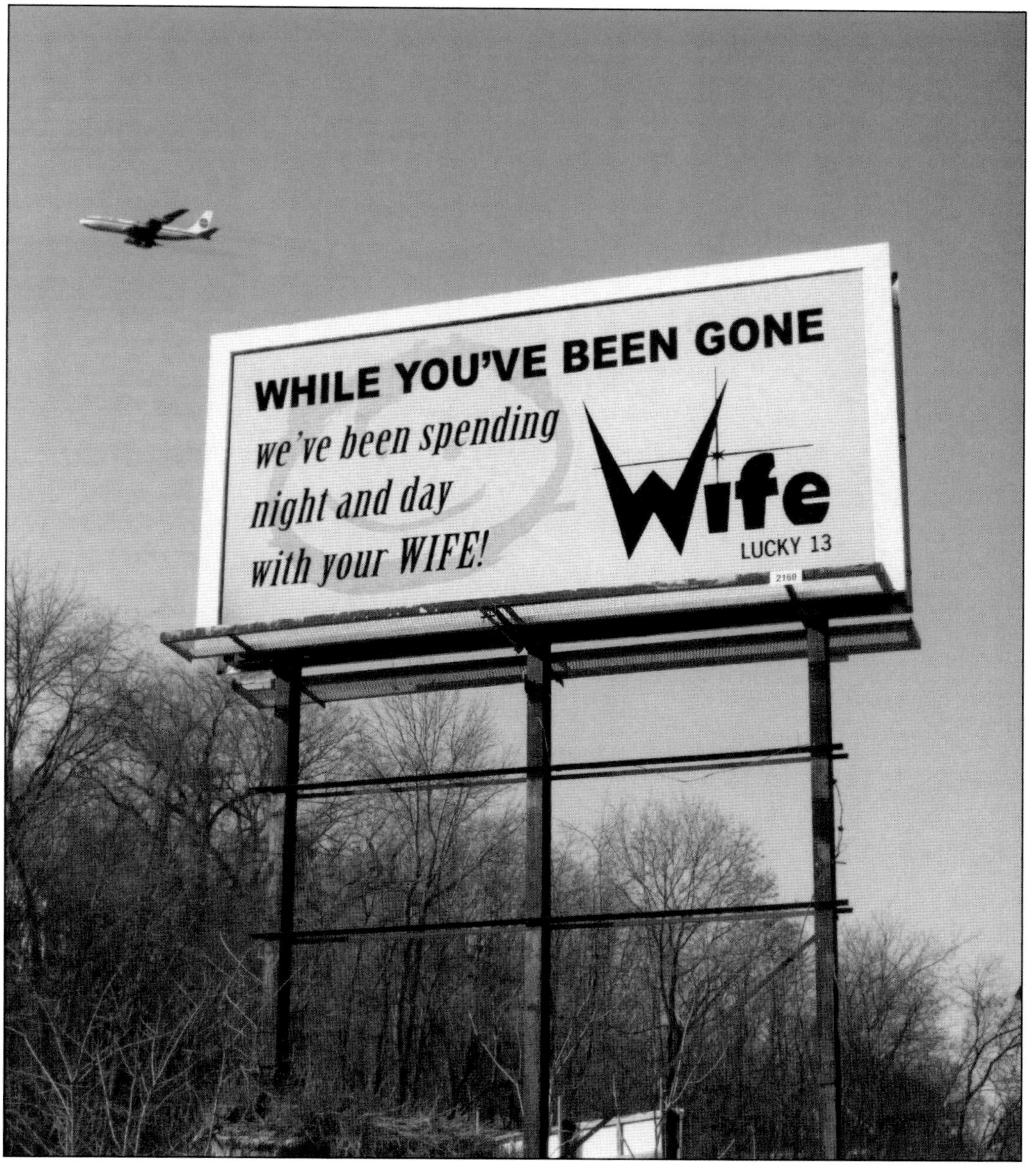

This WIFE billboard had a counterpart in Portland, Oregon, at another Don Burden station that read, "While you've been gone we've been KISN your wife!" (Composite photo.)

There was one promotion that WIFE had that everyone in Indianapolis remembers. It was a billboard out by the airport. —Al Stone

And when you went by it and you saw it—I can remember the logo, it was a big WIFE logo. And it said, "While you've been gone, we've been spending night and day with your WIFE!" And we used to get more feedback from that than you can imagine. —Jack Marsella

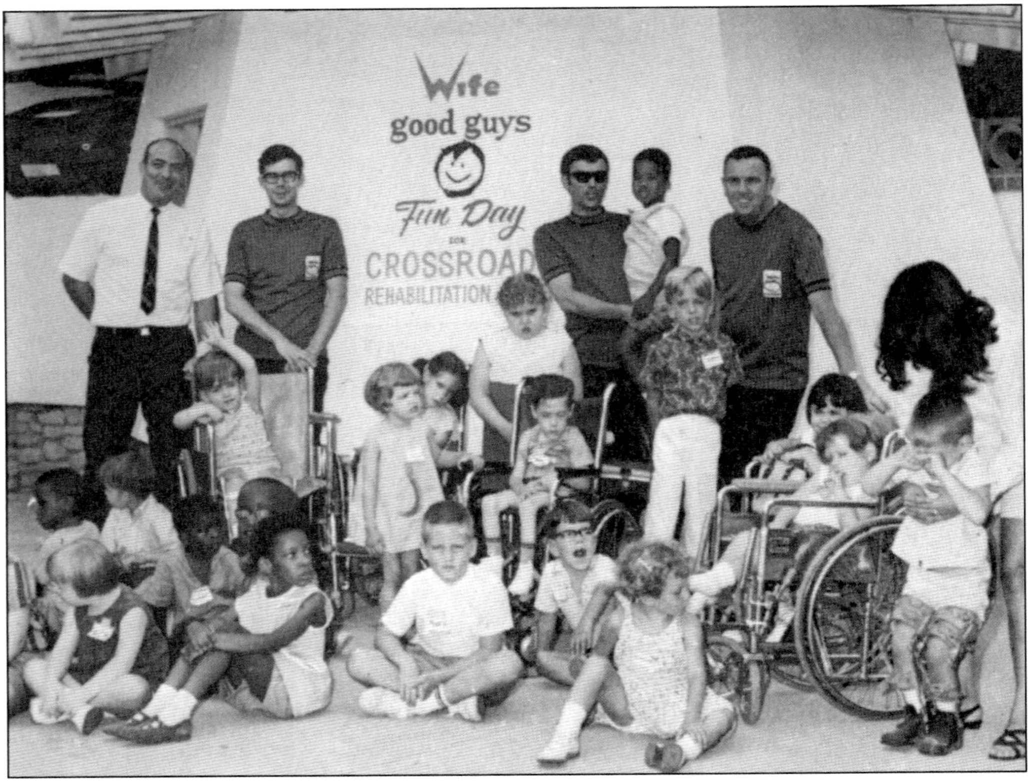

The WIFE Good Guys were everywhere in Indianapolis.

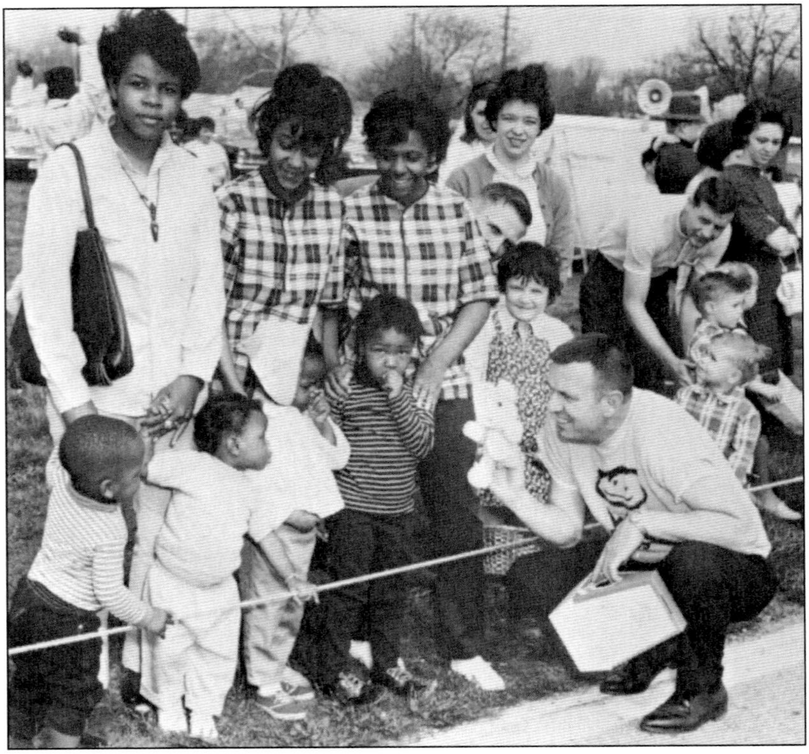

Everything was bigger than life. Everything we did was done to be bigger than life." —Roger W. Morgan

Well, the deal was to get more people listening to our station. And to do that, we had to give away things, I thought. And we gave away a lot of things. Homes and cars . . . and we weren't paying for it, you know. —Bob Kiley

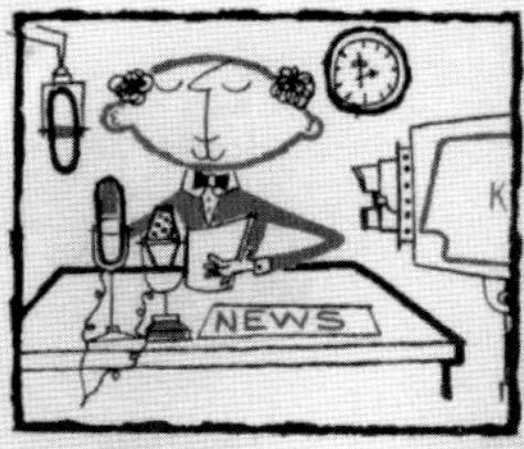

WIFE INDIANAPOLIS, IND.

special report

5,000/1,000w 1310kc IND

Indianapolis' favorite . . . .

WIFE MEASURES 37-45-55

The MOST VOLUPTUOUS figure in town!

--- Particularly when you're building the largest share of radio audience in one of the nation's greatest metropolises:

AM Drivetime - 37
Daytime - 45
PM Drivetime - 55

HOW WELL-STACKED is WIFE in competition?

6-9AM

Stations	Share	Listeners	CPM
WIFE	37	42,000	$1.17
A	21	24,400	1.11
B	20	22,900	1.54
C	16	18,400	1.52
D	6	7,200	2.78

(cont'd)

a service of H-R Television, Inc. Representatives, Inc.

Don Burden's choice of the WIFE call letters was calculated to be provocative.

And that was kind of a cool thing to be known as wife: WIFE. Where the call letters stood for something. WIFE—your wife. You know, we're tuned into your WIFE day and night. You know, whatever. I mean, that was a cool thing. —Mike "The Animal" O'Brien

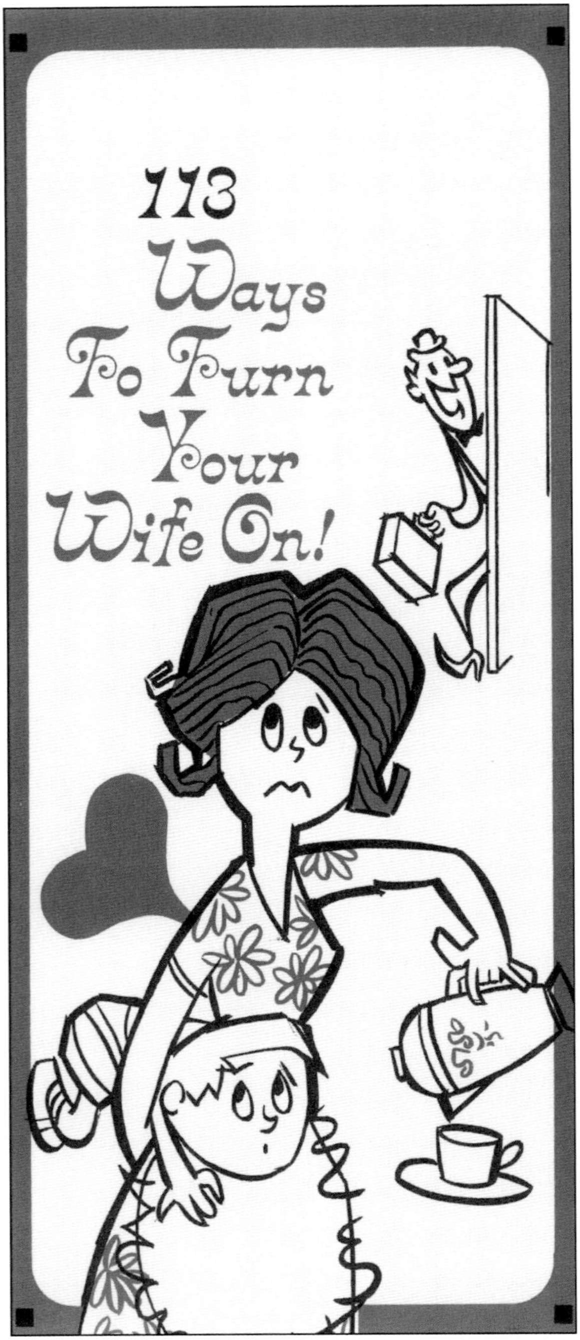

Lucky, because we had contesting going on. You could win color television sets. And color television sets back then was "oh, my goodness!" You know, nobody had that. And we gave away 113 color TV sets. —Pat Kelly

Don Burden also took advantage of the 1310 location on the dial, creating "Lucky 13."

They took advantage of every opportunity to imbed their call letters, which obviously were unique call letters, and their place on the dial in peoples' minds. Well, when you said you were lucky 13, they knew that. And they oriented the 13 to where it must be on the dial. So everything, the jingles were "lucky 13," the contests were "lucky 13," you had 13 prizes, you had things hidden at 13 locations. If anything would be oriented to the number 13, it was because that's where WIFE was on the dial: 1310. —Reb Porter

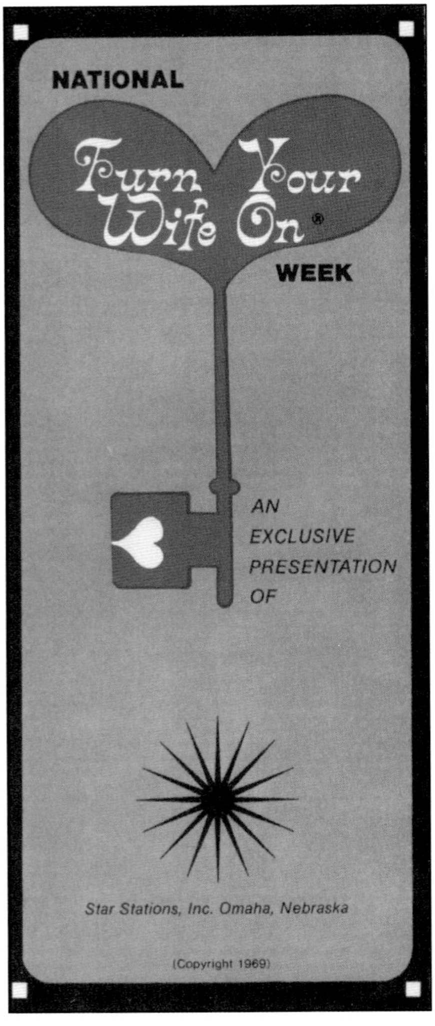

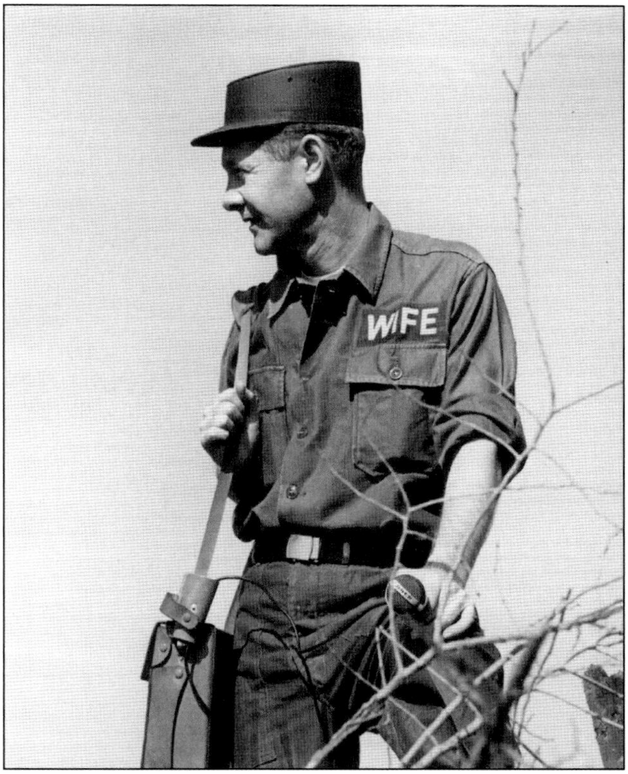

WIFE newsman Bill Donnella was assigned to cover one of the most controversial engagements of the 20th century, the Vietnam War. In 1966, he was the first Indiana newsman to visit the American military front. He said at the time, "This trip and my talks with servicemen from Indiana and throughout the nation gave me a new insight into the seriousness of this war."

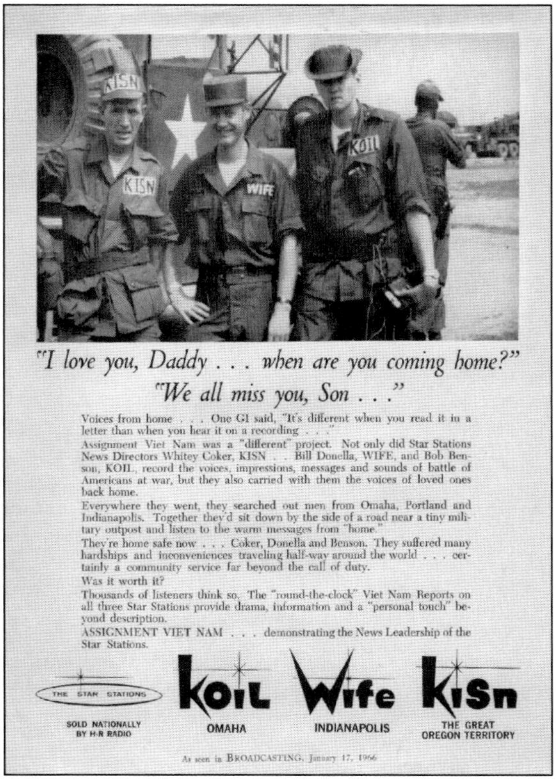

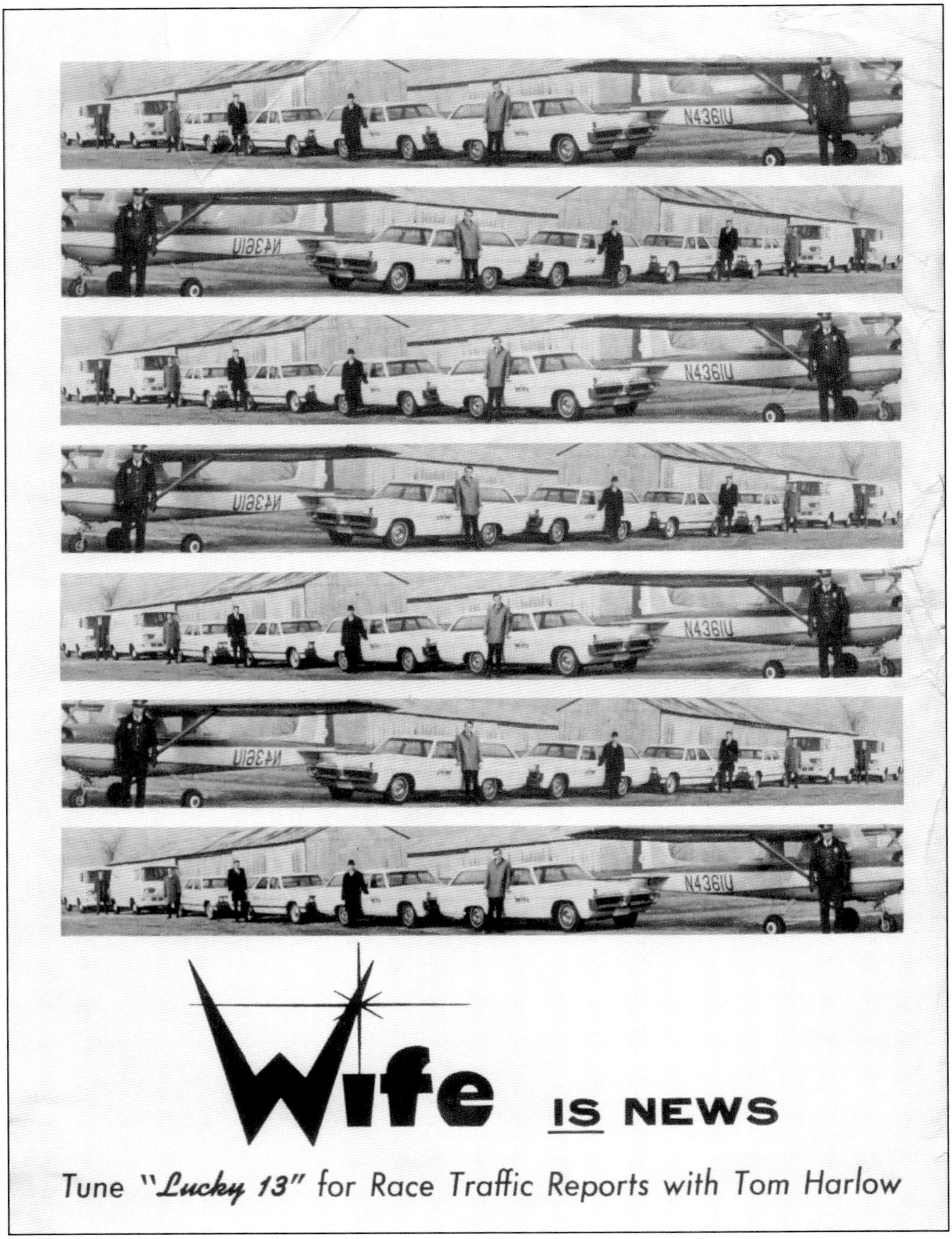

Operation Airwatch, with Lt. Tom Harlow offering traffic reports from an airplane every afternoon, was the first of its kind in Indianapolis. According to WIFE promotional materials, the station took it even further: "WIFE was instrumental in setting up an organization referred to as the 'Total Information News Network,' through which voice reports of any major new happening almost anywhere in the world can be heard reported from the scene within moments."

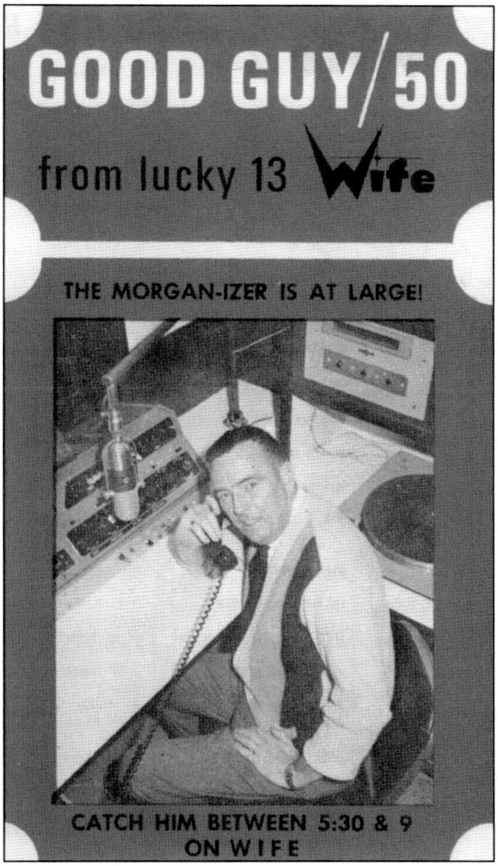

Robert W. Morgan was a very popular West Coast disc jockey. He was so well liked that Don Burden came up with his own character, Roger W. Morgan, for his Portland station KISN. That, in turn, was so popular he cloned the concept and had two more Roger W. Morgans for his Omaha and Indianapolis stations. When the Indianapolis Morgan (at right) resigned, Burden brought in the original Roger W. Morgan from the Portland station to be the new Indianapolis Roger W. Morgan (below). This led to untold confusion among the listeners.

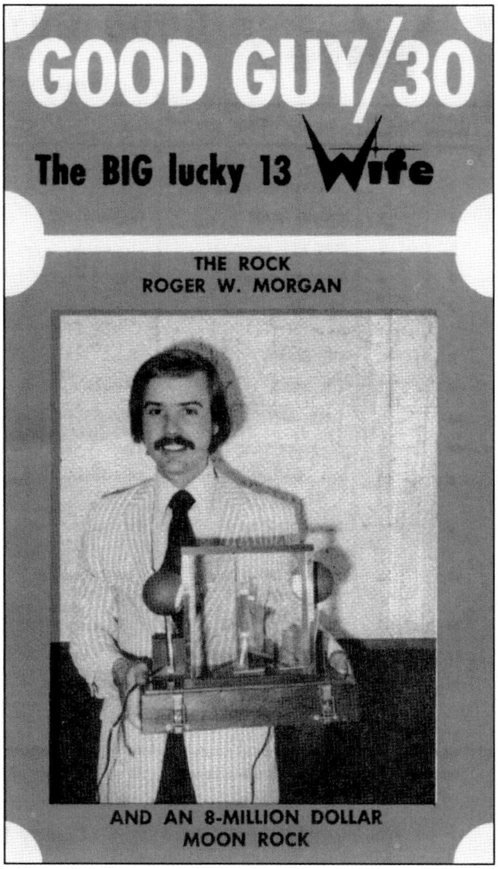

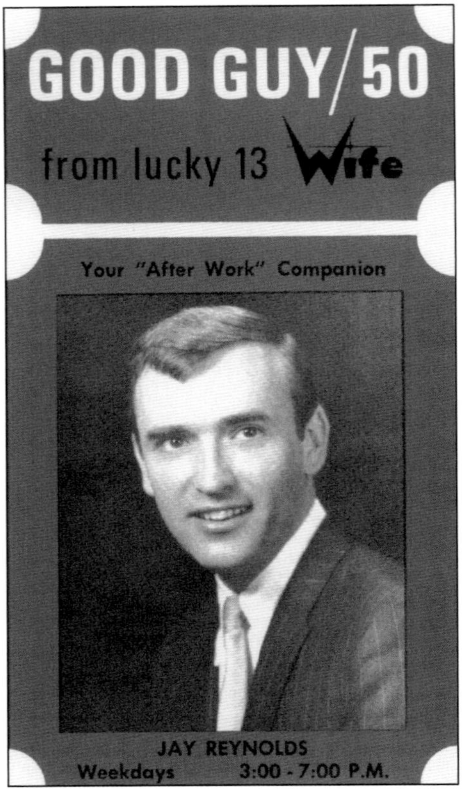

No one epitomized the WIFE on-air sound more than Jay Reynolds.

The pacing was very tight, very clean, not a lot of extra words were used. They actually speeded up their records so they could get more music in. The disc jockeys had lots of energy. —Art Vuolo Jr.

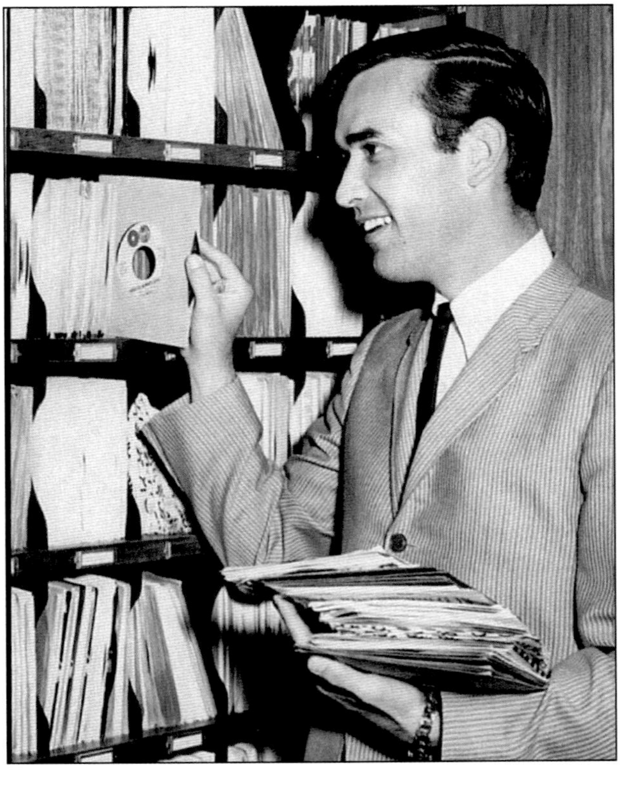

My favorite deejay became Jay Reynolds. The "Jay-bird" in the catbird seat, every afternoon three to seven on WIFE. He was absolutely incredible. —Scott Shannon

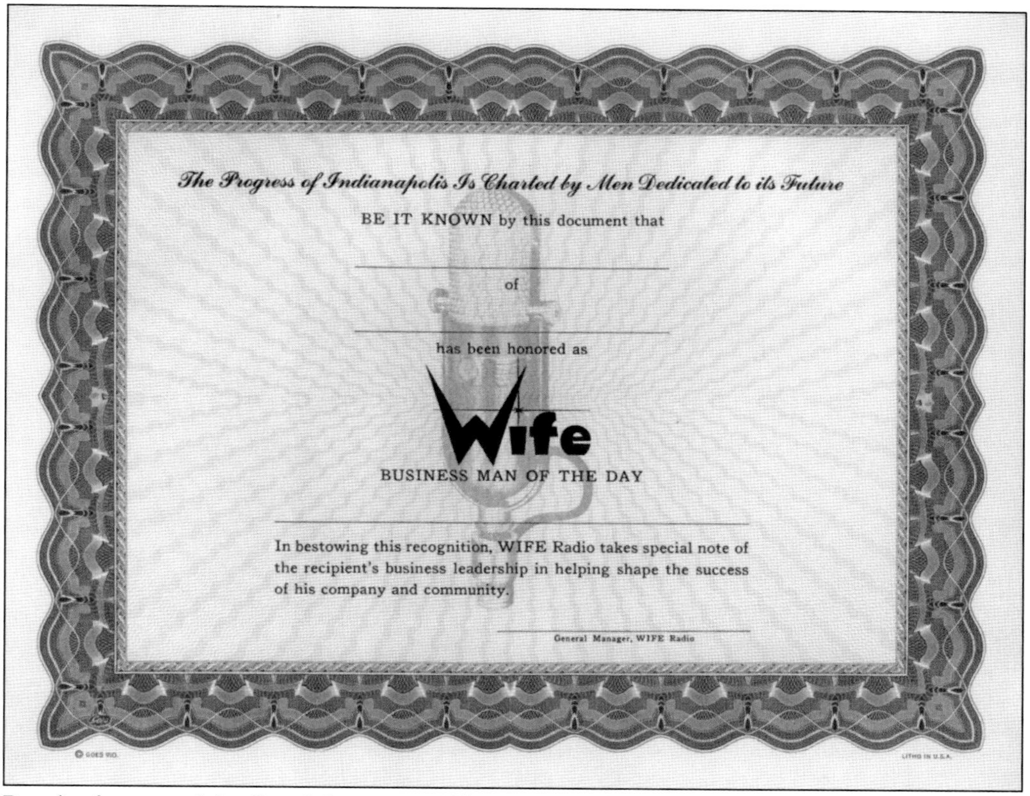

The Progress of Indianapolis Is Charted by Men Dedicated to its Future

BE IT KNOWN by this document that

of

has been honored as

WIFE

BUSINESS MAN OF THE DAY

In bestowing this recognition, WIFE Radio takes special note of the recipient's business leadership in helping shape the success of his company and community.

General Manager, WIFE Radio

Regular features of the Good Guys were "Clergyman of the Day" and "Businessman of the Day." (Below, Composite photo.)

We did a WIFE Good Guy "Businessman of the Day." And we would salute a businessperson within the community every hour, on the hour, from early morning throughout the course of each day.
—Jack Marsella

The "Businessman of the Day" had one primary purpose and that was to convince reluctant advertisers that there was power in radio. Because the moment somebody became businessman of the day, his phone rang off the wall.
—Roger W. Morgan

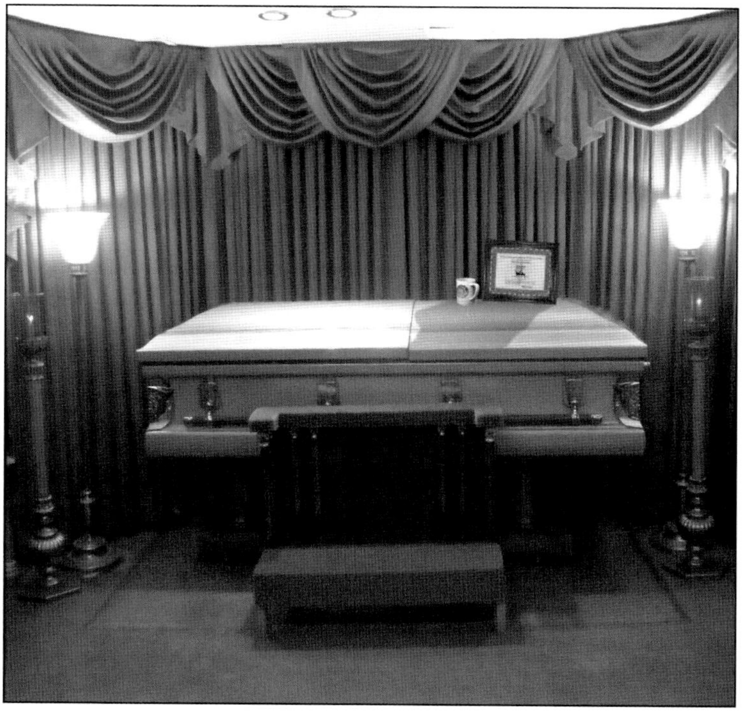

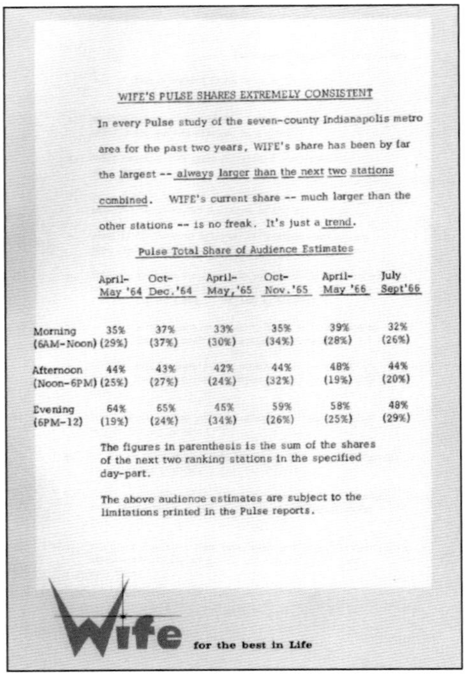

WIFE'S PULSE SHARES EXTREMELY CONSISTENT

In every Pulse study of the seven-county Indianapolis metro area for the past two years, WIFE's share has been by far the largest -- always larger than the next two stations combined. WIFE's current share -- much larger than the other stations -- is no freak. It's just a trend.

Pulse Total Share of Audience Estimates

	April-May '64	Oct-Dec.'64	April-May,'65	Oct-Nov.'65	April-May '66	July Sept'66
Morning (6AM-Noon)	35% (29%)	37% (37%)	33% (30%)	35% (34%)	39% (28%)	32% (26%)
Afternoon (Noon-6PM)	44% (25%)	43% (27%)	42% (24%)	44% (32%)	48% (19%)	44% (20%)
Evening (6PM-12)	64% (19%)	65% (24%)	45% (34%)	59% (26%)	58% (25%)	48% (29%)

The figures in parenthesis is the sum of the shares of the next two ranking stations in the specified day-part.

The above audience estimates are subject to the limitations printed in the Pulse reports.

Wife for the best in Life

Ratings are everything in broadcasting.

I think the feeling of having pressure on you at all times gets tiresome. And we had guys that lasted less than a week. —Bill Donnella

The ratings were phenomenal, just phenomenal. They were so high that they were unbelievable. —Pat Kelly

During the Sixties when WIFE was the dominant station, the competition was basically zero. "The Emperor" Joe Light was quite a personality and was on the air from seven till midnight. And at one point in that period of time, he had 99 percent of the audience at night. During the rest of the air shifts, if we didn't have more than 40 percent of the audience—preferably 50 percent to 60 percent of the audience—our jobs might have been in jeopardy. —Reb Porter

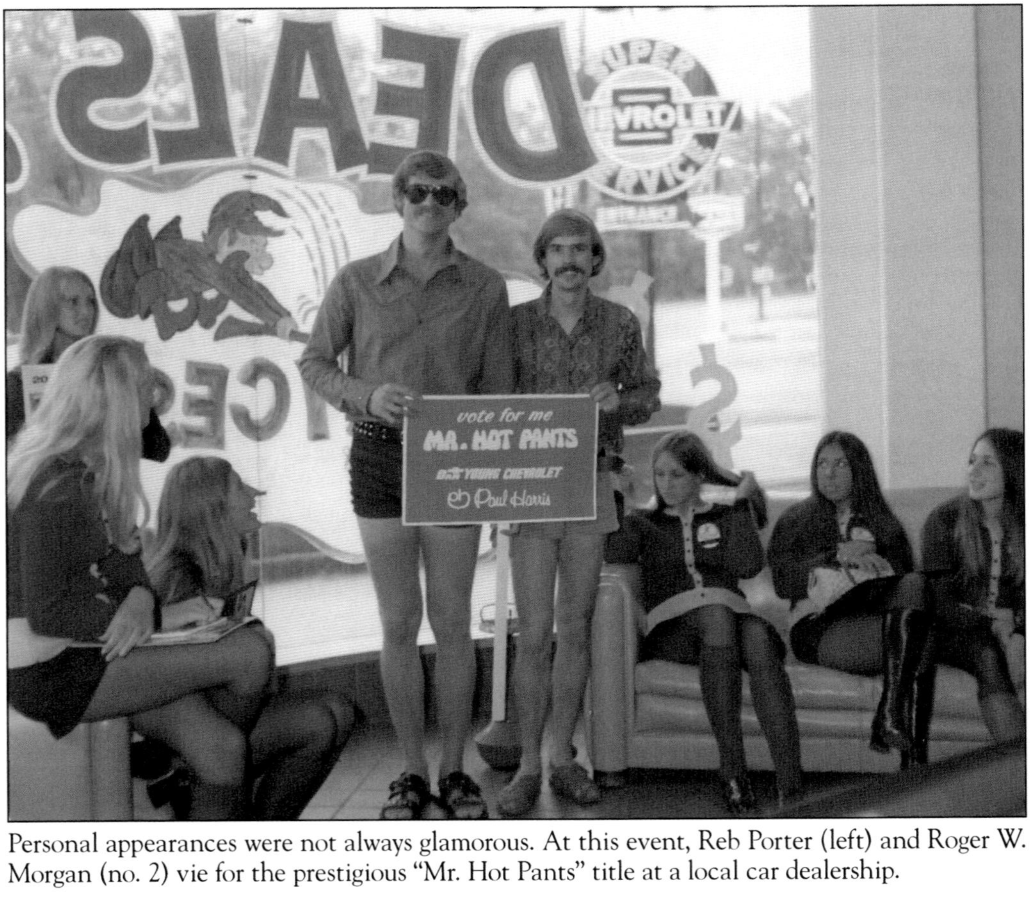

Personal appearances were not always glamorous. At this event, Reb Porter (left) and Roger W. Morgan (no. 2) vie for the prestigious "Mr. Hot Pants" title at a local car dealership.

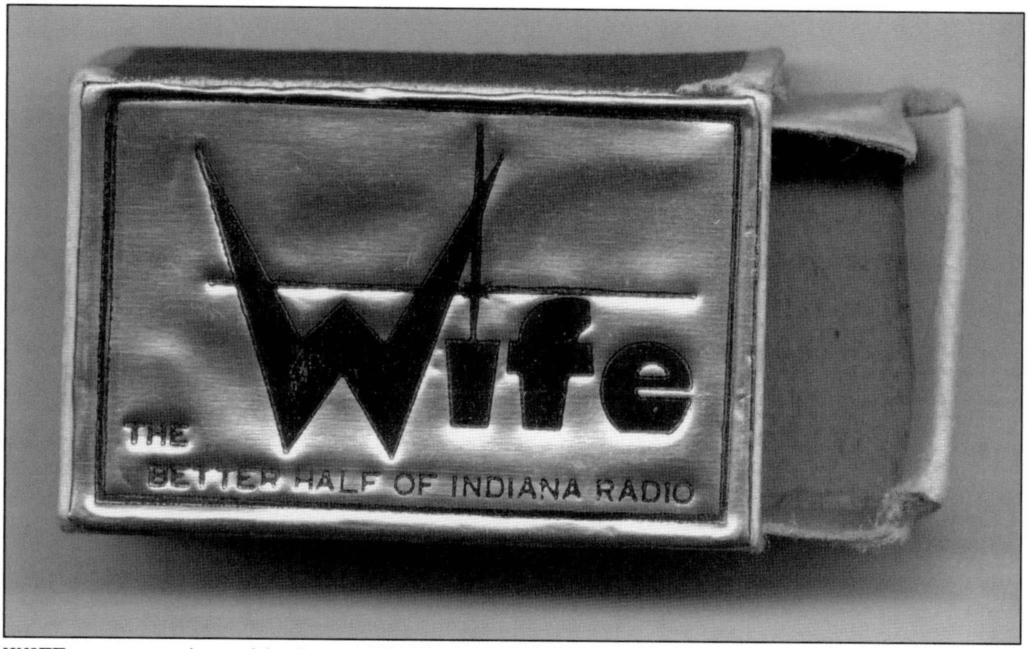

WIFE gave away lots of freebies in the 1960s. This is a vintage matchbox from the station with the inscription "The Better Half of Indiana Radio." WIFE commissioned Tom Merriman, who had produced all of the station's jingles throughout 1964, to produce an extended song that was pressed into a promotional 45 (below).

The ratings are the most important thing in radio. And jingles can help get a station in the running simply by burning into a person's mind a certain melody or a certain phrase about a radio station.
—Steve Cooper

WIFE had probably some of the best jingles this town has ever heard, which really transitioned things nicely.
—Art Vuolo Jr.

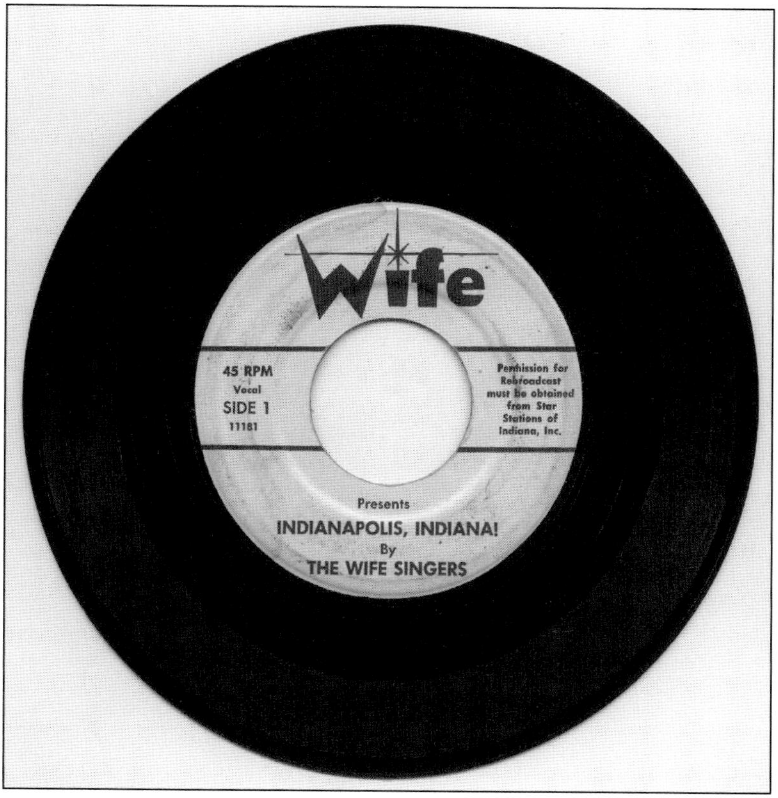

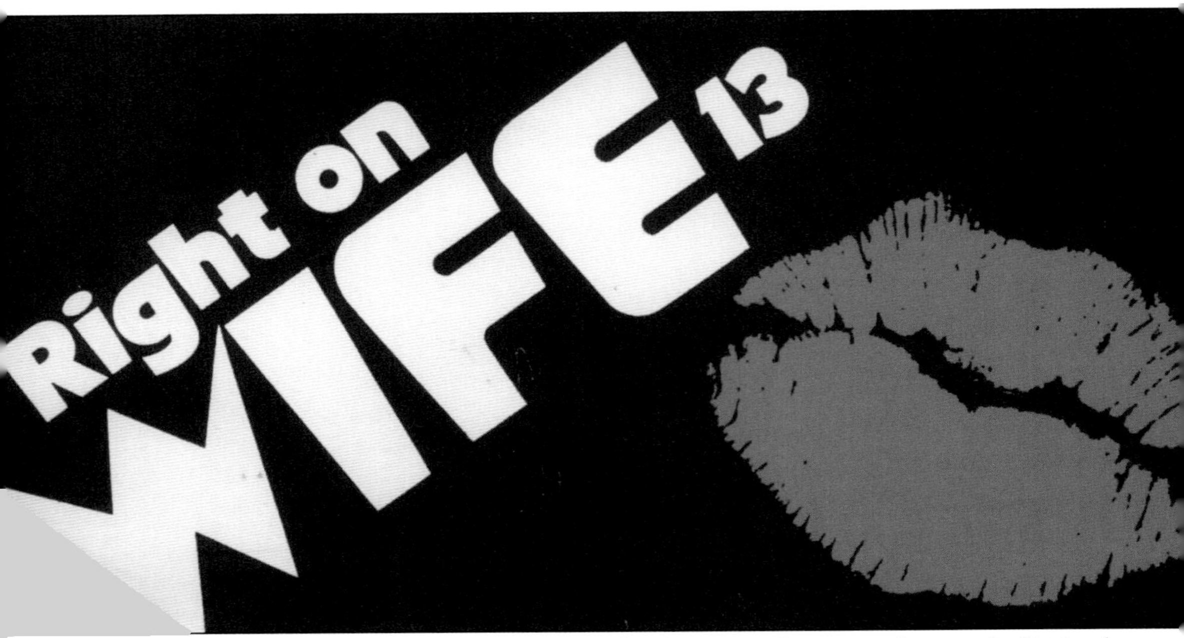

In 1973, George Johns came to Indianapolis to be the national program director for Fairbanks Broadcasting, working directly under Jim Hilliard. He recalls when he stepped off the plane and was driving to the station, he saw a billboard with the "Right on WIFE" slogan. He knew right away that WIFE was dead. No one cool had used the term "right on!" for years. This was a radio station that was out of touch with its audience, and WNAP was going to capitalize on that.

Three

FM STRIKES BACK WITH WNAP

Fairbanks Broadcasting, which owned WIBC-AM and WIBC-FM, was mired in disappointment in the mid-1960s. Ratings for Bouncin' Bill Baker were down, and revenues were suffering. Owner Richard M. Fairbanks was sitting on his two Indianapolis stations and just could not seem to capture the magic WIFE had stolen from him.

In 1968, Fairbanks brought in Jim Hilliard in hopes of turning ratings around. While at WFIL in Philadelphia, Hilliard had become the first program director in the country to create counterprogramming against Bill Drake's formula, turning WFIL into the market's no. 1 station. Hilliard then went to Fairbanks Broadcasting and turned that company, Indianapolis radio, and broadcasting in general on end.

Against all odds, WNAP became the nation's first successful FM-only rock station, which created a genuine radio war in the market. It was AM radio versus fledgling FM radio, album cuts versus 45s, and the Good Guys versus the "bad boys." In 1973, Hilliard brought in George Johns, an old colleague from Hilliard's days working at a station in Winnipeg, Manitoba, to be national program director. Hilliard and Johns were visionaries. According to John Blair & Company, Fairbanks's national marketing firm, WNAP was "the highest billing FM station" on their list.

The real money on the market was on two stations. It was at WFBM, and the other one was at WIFE, when I came back. And I said, "Mr. Fairbanks, we can't attack one, we have to attack them both. If you're going to get into a war, you're going to get dirty. So let's go get dirty." —Jim Hilliard

WIBC-FM was a classical music station, and it was run on 24-hour tapes. One large reel of tape, and it would go real slow. And nobody paid any attention to it. And either one of the engineers had forgotten, or the tape broke, I'm not sure . . . but it had probably been off the air for 8, 10, maybe 12 hours and nobody had received a call. So they knew there wasn't anybody listening. And they knew they wanted to do something with it. —Cris Conner

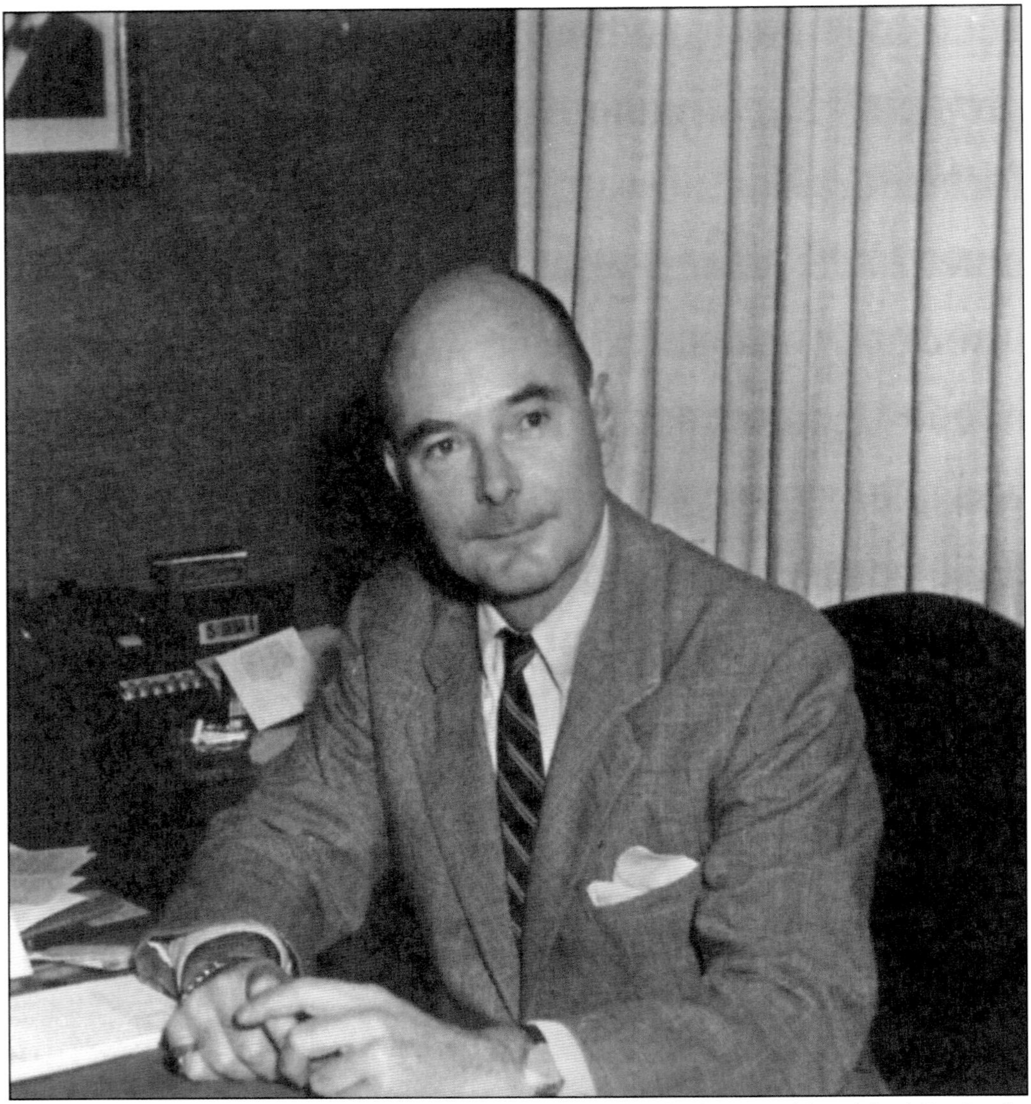

Don Burden owned Star Stations, and Richard M. Fairbanks (pictured here) had Fairbanks Broadcasting.

Fairbanks Broadcasting was owned exclusively by Richard Fairbanks, the grandson of Teddy Roosevelt's vice president, Charles Warren Fairbanks, for whom Fairbanks, Alaska, was named. Dick was a fiercely proud radio magnate operating 20 radio stations, a television station in Atlanta, and numerous cable television systems. He was a journalist, his family owning the Indianapolis News *for some 50 years.* —Cris Conner

Jim Hilliard had been an on-air personality in Indianapolis in the mid-1960s under the guise of "Jimmy Darin." In 1968, the first meeting of Hilliard and Cris Conner happened after-hours. Owner Richard M. Fairbanks was taking Hilliard on a tour of the station. Conner was on the air with the evening shift at the back of the building in the WNAP studio. Hilliard stood in the doorway and saw this kid, Cris, wearing a skin-tight T-shirt yelling and banging a gong. Hilliard's reaction was "I'm going to keep this guy."

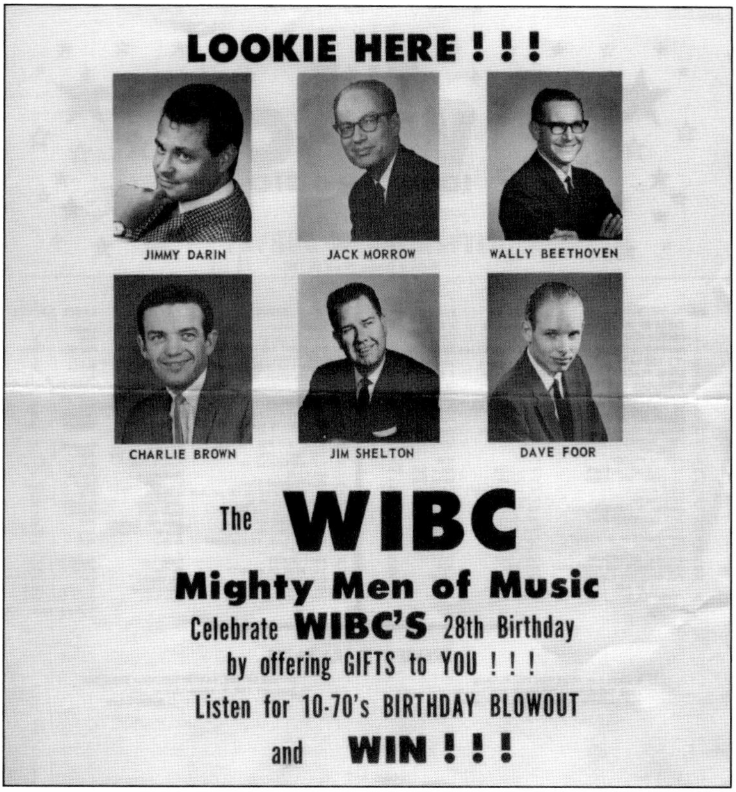

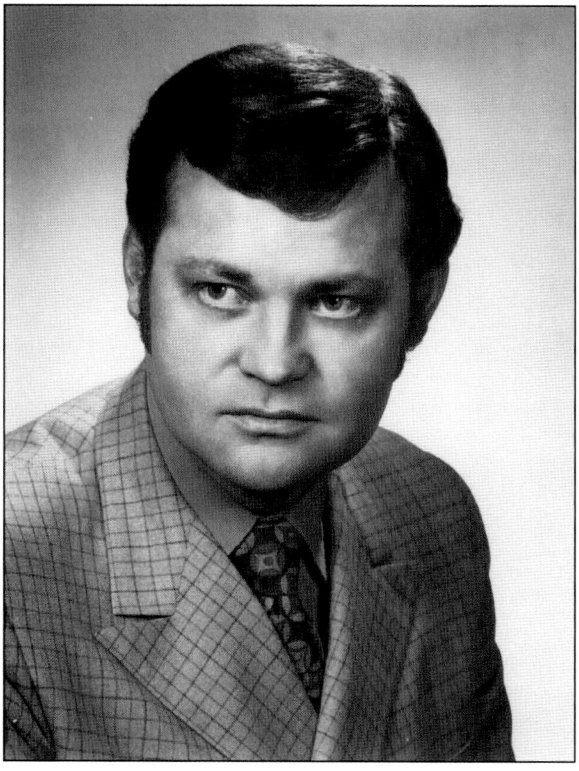

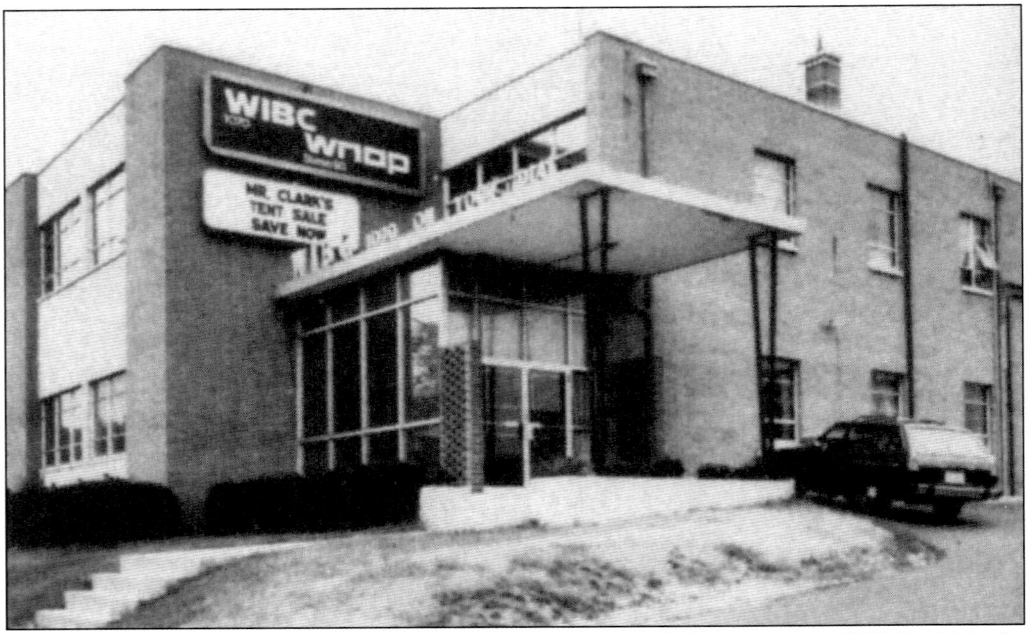

This is the legendary "Two-Story Brick" building at 2835 North Illinois Street. WIBC-AM had a nice studio at the front of the building, while the WNAP studio was at the very back in a room the size of a closet.

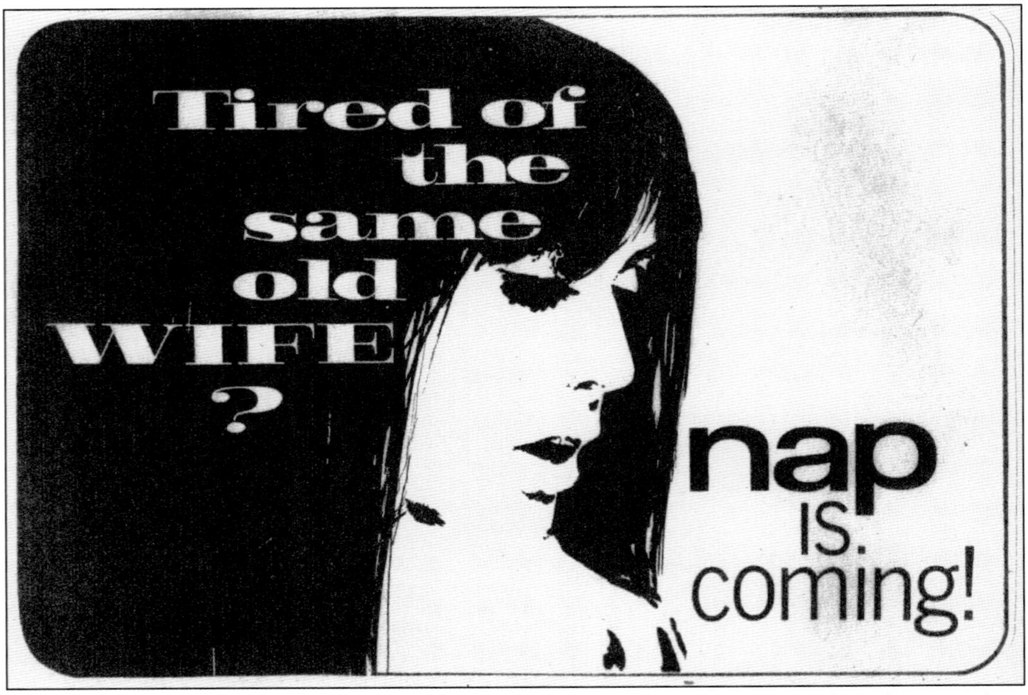

In an obvious dig at WIFE, WNAP used the city's counterculture to build a brand centered on radical, edge-of-your-seat broadcasting. Along those lines was the selection of the call letters for WNAP, which were derived from the longstanding nickname for Indianapolis, "Naptown." No one is quite sure where the nickname came from, although the author believes it is from the "nap" in Indianapolis, but most agree it has a derogatory connotation.

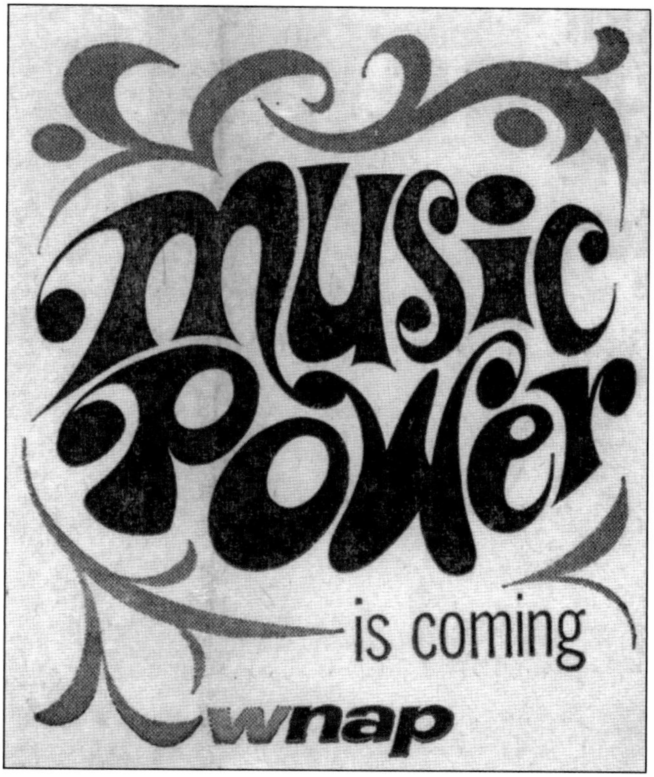

These are early promotional fliers featuring art styles of the time by artist/writer Jeff Leiendecker. The two-finger peace sign was held up with his left hand while using the right to create the drawing.

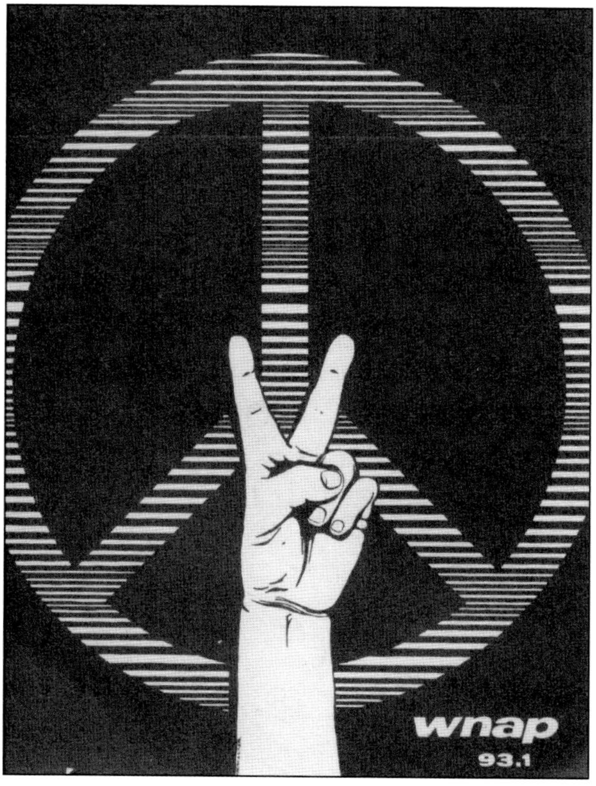

WNAP, INDIANAPOLIS

SPECIAL INTRODUCTORM RATE OFFER

RADIO STATION WNAP WILL BEGIN OPERATIONS SOON, AT 93.1 MEGACYCLES ON FM. COINCIDENTAL WITH THE START OF OPERATIONS WILL BE THE FIRST ABFILIATION BY THE AMERICAN CONTEMPORARY NETWORK, WITH AN FM STATION. THE STAFF OF THE ALL-NEW STATION IS VIRTUALLY COMPLETE, AND WE ARE AWAITING THE INSTALLATION OF A NEW TRANSMITTER AND ANTENNA ON THE MERCHANTS NATIONAL BANK BUILDING IN DOWNTOWN INDIANAPOLIS.

TO INAUGURATE THE NEW BROADCAST SERVICE, WNAP IS OFFERING ADVERTISERS AN OPPORTUNITY TO USE THE FACILITIES AT A GREATLY REDUCED PRICE. THE PLAN, BASED ON AN EIGHT-WEEK SCHEDULE, IS AS FOLLOWS:

1. BUY THE FIRST FOUR-WEEK SCHEDULE AT REGULAR RATES.
2. BUY THE SAME SCHEDULE THE FOLLOWING FOUR WEEKS, AND RECEIVE ONE OF TWO BONUS ARRANGEMENTS;
   A. GET TWICE THE NUMBER OF ANNOUNCEMENTS FOR THE SAME DOLLAR FIGURE.
   OR B. GET THE SAME NUMBER OF ANNOUNCEMENTS AT ONE-HALF THE PUBLISHED RATE-CARD PRICE.

THIS IS AN OUTSTANDING OPPORTUNITY TO BE IN ON THE GROUND FLOOR OF AN EXCITING NEW RADIO STATION IN INDIANAPOLIS---ONE WHICH WILL HAVE ALL THE PROMOTION NECESSARY TO PUT IT ON THE RADIO MAP QUICKLY IN OUR CITY.

NOW IS THE TIME TO SIGN.

Initially, selling airtime for commercials was tough. This "cheat sheet" was used by the WNAP sales staff to drum up interest. It was an uphill climb: the station was on the FM band with very few cars having FM receivers, and it was unproven with no ratings track record.

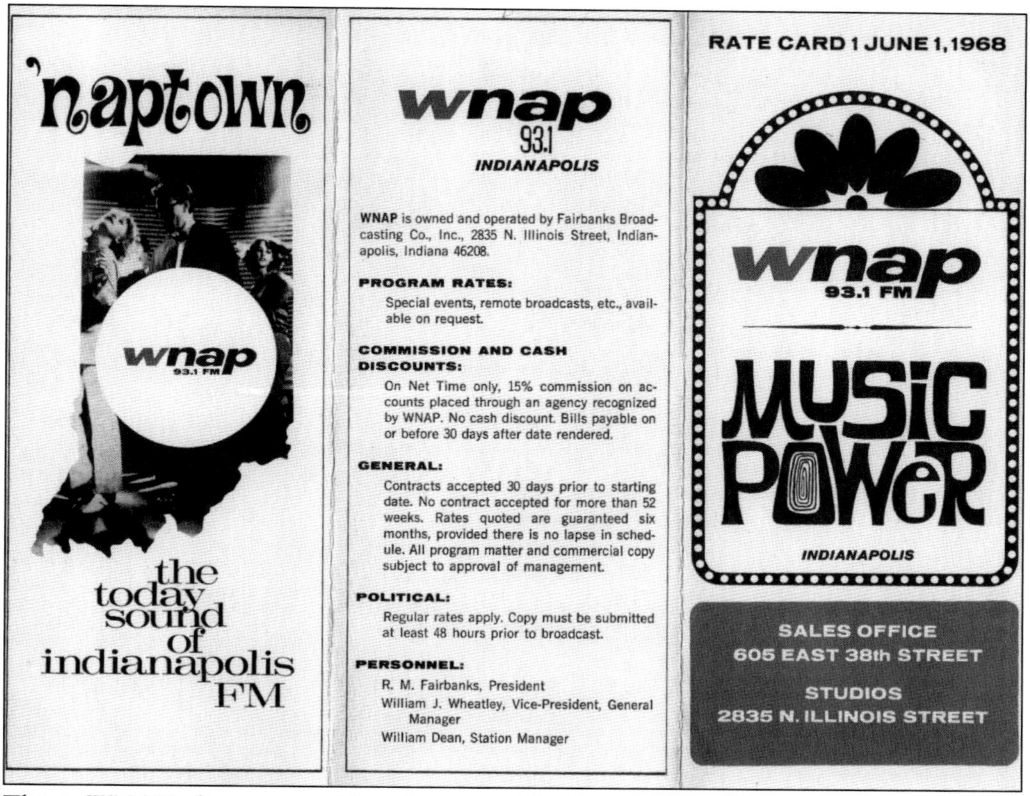

This is WNAP's first rate card for advertising. The June 1, 1968, date is of interest since the station did not sign on until July 22, 1968. These ad rates were considered dirt cheap, even for those days. With each on-air staff member being paid $125 per week and working in a "studio" containing a patchwork of unused WIBC equipment in a closet at the back of the building, Fairbanks Broadcasting had nothing to lose and a blueprint that might just succeed.

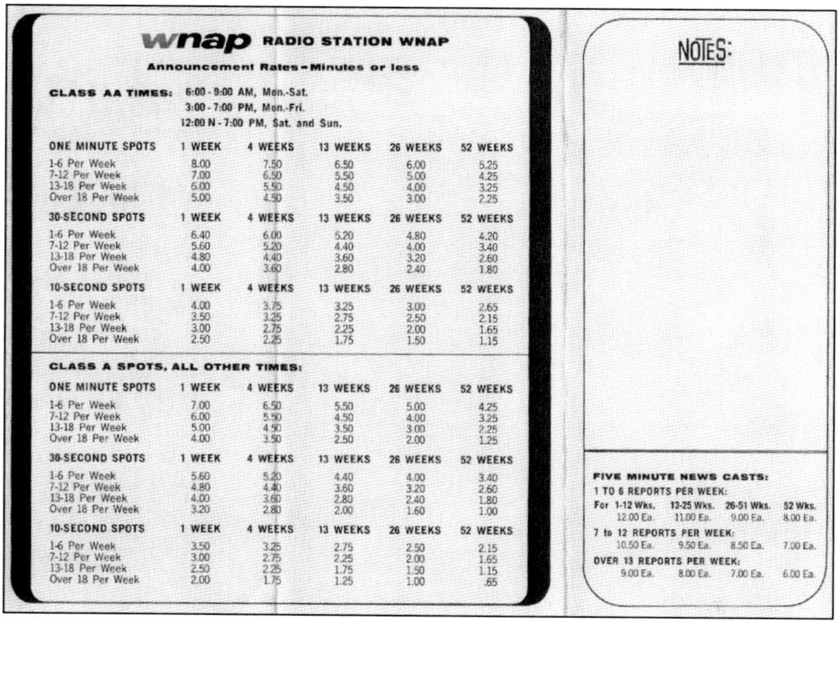

wnap RADIO STATION WNAP

Announcement Rates—Minutes or less

CLASS AA TIMES: 6:00 - 9:00 AM, Mon.-Sat.
3:00 - 7:00 PM, Mon.-Fri.
12:00 N - 7:00 PM, Sat. and Sun.

ONE MINUTE SPOTS	1 WEEK	4 WEEKS	13 WEEKS	26 WEEKS	52 WEEKS
1-6 Per Week	8.00	7.50	6.50	6.00	5.25
7-12 Per Week	7.00	6.50	5.50	5.00	4.25
13-18 Per Week	6.00	5.50	4.50	4.00	3.25
Over 18 Per Week	5.00	4.50	3.50	3.00	2.25
30-SECOND SPOTS	**1 WEEK**	**4 WEEKS**	**13 WEEKS**	**26 WEEKS**	**52 WEEKS**
1-6 Per Week	6.40	6.00	5.20	4.80	4.20
7-12 Per Week	5.60	5.20	4.40	4.00	3.40
13-18 Per Week	4.80	4.40	3.60	3.20	2.60
Over 18 Per Week	4.00	3.60	2.80	2.40	1.80
10-SECOND SPOTS	**1 WEEK**	**4 WEEKS**	**13 WEEKS**	**26 WEEKS**	**52 WEEKS**
1-6 Per Week	4.00	3.75	3.25	3.00	2.65
7-12 Per Week	3.50	3.25	2.75	2.50	2.15
13-18 Per Week	3.00	2.75	2.25	2.00	1.65
Over 18 Per Week	2.50	2.25	1.75	1.50	1.15

CLASS A SPOTS, ALL OTHER TIMES:

ONE MINUTE SPOTS	1 WEEK	4 WEEKS	13 WEEKS	26 WEEKS	52 WEEKS
1-6 Per Week	7.00	6.50	5.50	5.00	4.25
7-12 Per Week	6.00	5.50	4.50	4.00	3.25
13-18 Per Week	5.00	4.50	3.50	3.00	2.25
Over 18 Per Week	4.00	3.50	2.50	2.00	1.25
30-SECOND SPOTS	**1 WEEK**	**4 WEEKS**	**13 WEEKS**	**26 WEEKS**	**52 WEEKS**
1-6 Per Week	5.60	5.20	4.40	4.00	3.40
7-12 Per Week	4.80	4.40	3.60	3.20	2.60
13-18 Per Week	4.00	3.60	2.80	2.40	1.80
Over 18 Per Week	3.20	2.80	2.00	1.60	1.00
10-SECOND SPOTS	**1 WEEK**	**4 WEEKS**	**13 WEEKS**	**26 WEEKS**	**52 WEEKS**
1-6 Per Week	3.50	3.25	2.75	2.50	2.15
7-12 Per Week	3.00	2.75	2.25	2.00	1.65
13-18 Per Week	2.50	2.25	1.75	1.50	1.15
Over 18 Per Week	2.00	1.75	1.25	1.00	.65

NOTES:

FIVE MINUTE NEWS CASTS:

1 TO 6 REPORTS PER WEEK:				
For	1-12 Wks.	13-25 Wks.	26-51 Wks.	52 Wks.
	12.00 Ea.	11.00 Ea.	9.00 Ea.	8.00 Ea.
7 to 12 REPORTS PER WEEK:	10.50 Ea.	9.50 Ea.	8.50 Ea.	7.00 Ea.
OVER 13 REPORTS PER WEEK:	9.00 Ea.	8.00 Ea.	7.00 Ea.	6.00 Ea.

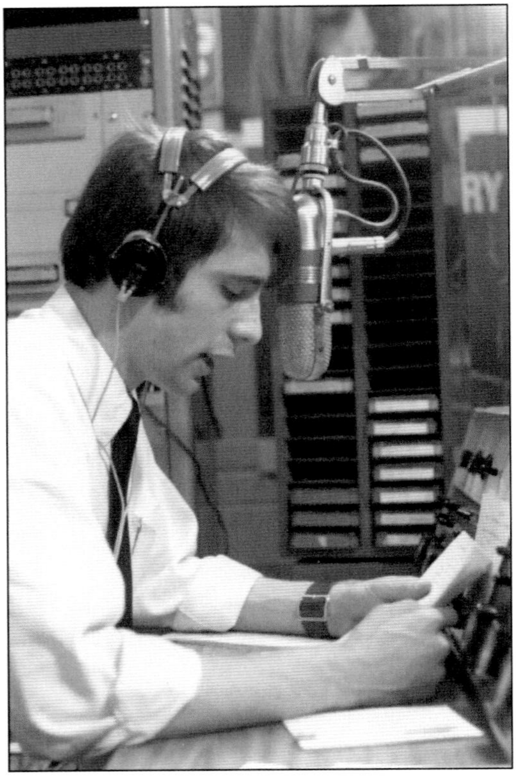

Al Stone is pictured in the original studio.

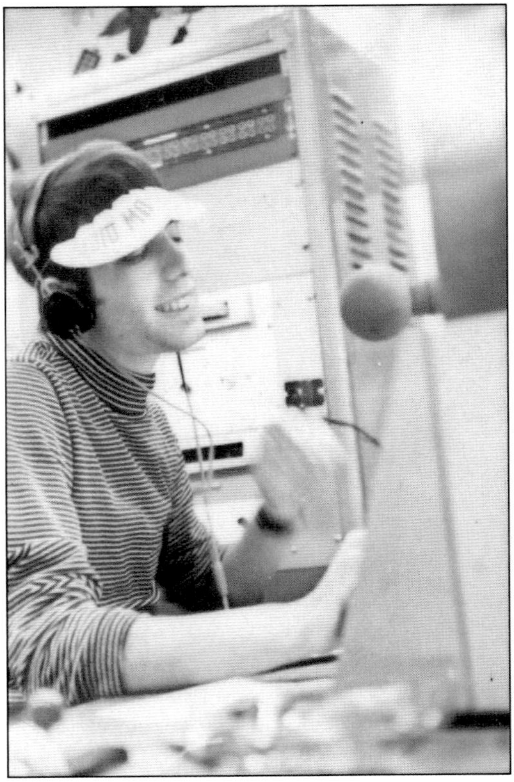

Jim Hanson and I had just come out from IU to teach in Indianapolis, and we were working at WIFE as nighttime and overnight news guys. And I got a call to go see the general manager at WIBC. And he hired us on the spot to take over the FM station and turn it into this real modern, hip rock station. —Al Stone

This c. 1969 photograph of WNAP staff features, from left to right, Cris Conner, Al Stone (seated), Jim Hansen, and Major Tom Lewis.

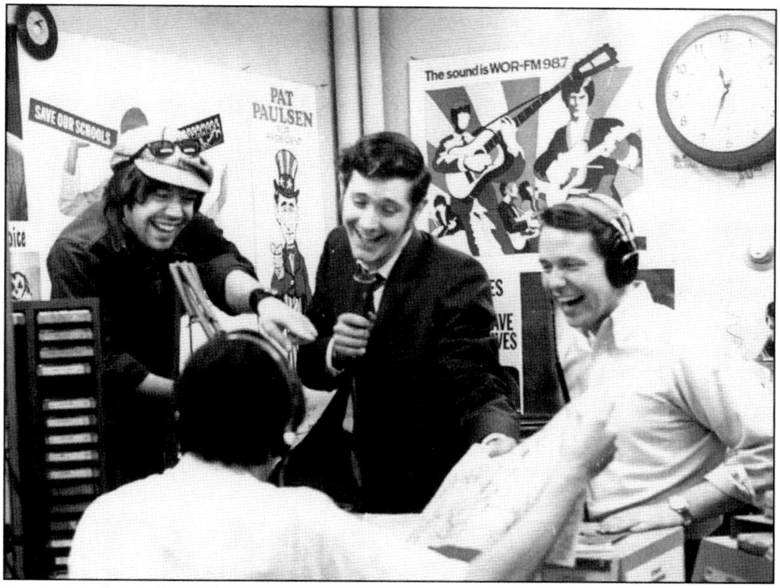

For the first three or four months, I don't know what WIFE did. I don't think they were overly concerned. I think probably the best thing we did was put those kids on the air, because even though they were good, I don't think they frightened WIFE. Because they were so disjointed. It was not a well-packaged product. Dare I say that without insulting anybody. —Jim Hilliard

At WNAP, they found five or six guys that had worked together, many of us in college. A couple of us had actually been on commercial radio before. And we were all available. And we came together, and we clicked. We liked each other. We loved what we were doing. We were too young and stupid to know that it couldn't be done. —John Gillis

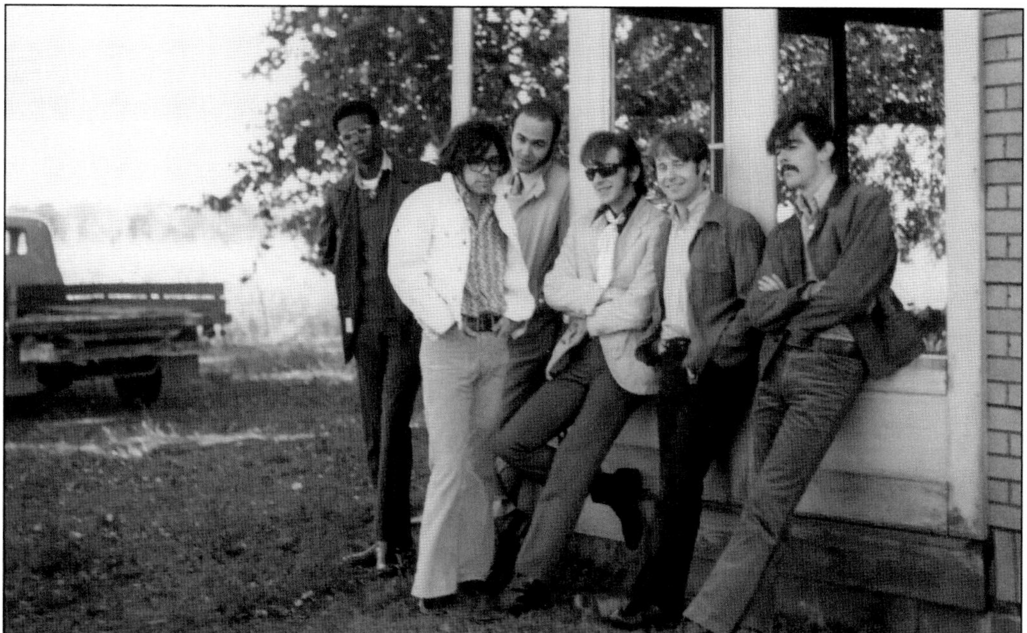

The WNAP staff pictured here around 1970 included, from left to right, Dave "Bad Results" Roosevelt, Cris Conner, "Big" John Gillis, Steve King, Major Tom Lewis, and Charlie Kendal.

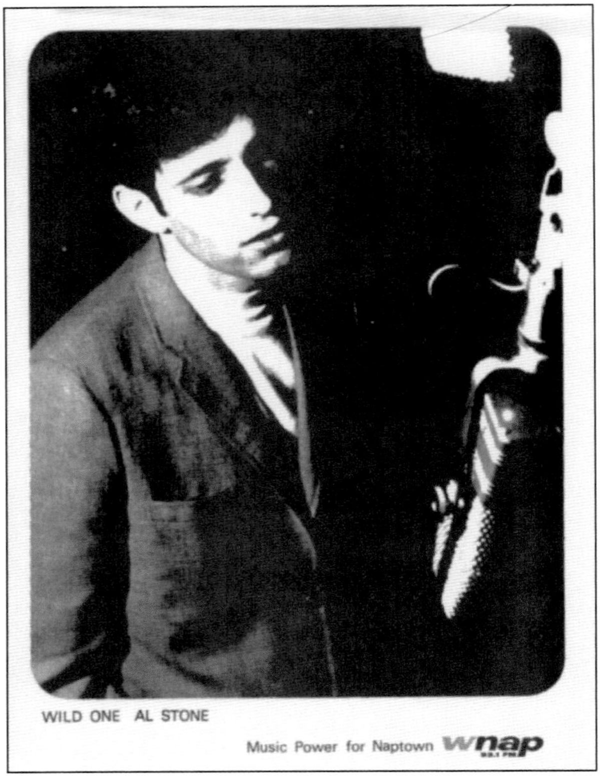

The original co–program directors for WNAP were Al Stone and Jim Hanson, who used the airname "Jasen."

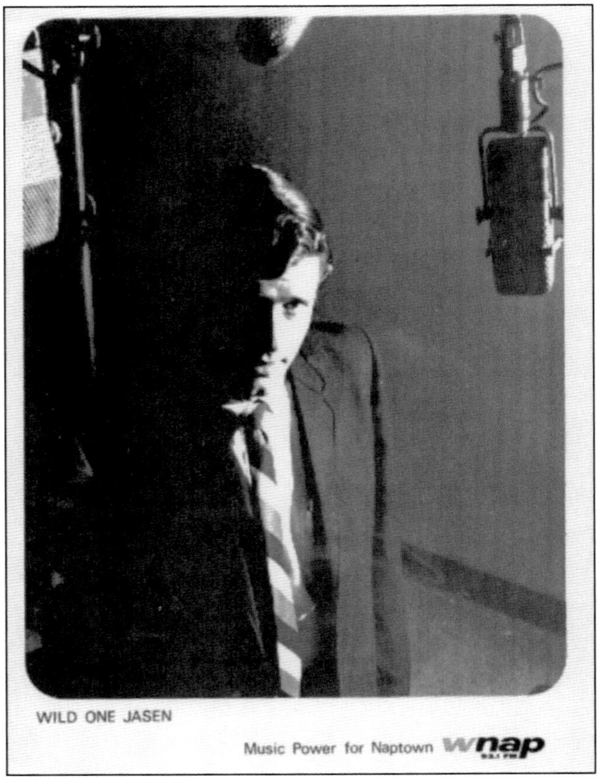

"Big" John Gillis, Rick Reinhart, and Cris Conner rounded out the original staff. Gillis was the first to sign WNAP on the air at 6:00 a.m. on July 22, 1968.

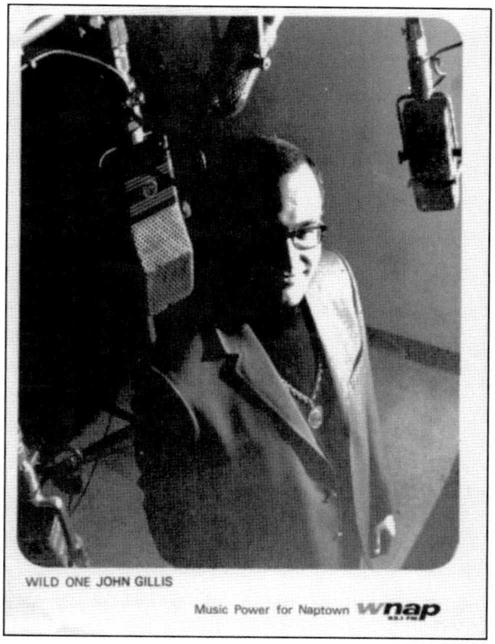

The first thing we did was play "The National Anthem." And it was the same old scratchy national anthem record by the Air Force Band that I played all my life in Sturgis, Michigan, or anywhere else. The first record we ever played was a song called "I Am the Muffin Man" by The World of Oz. —"Big" John Gillis

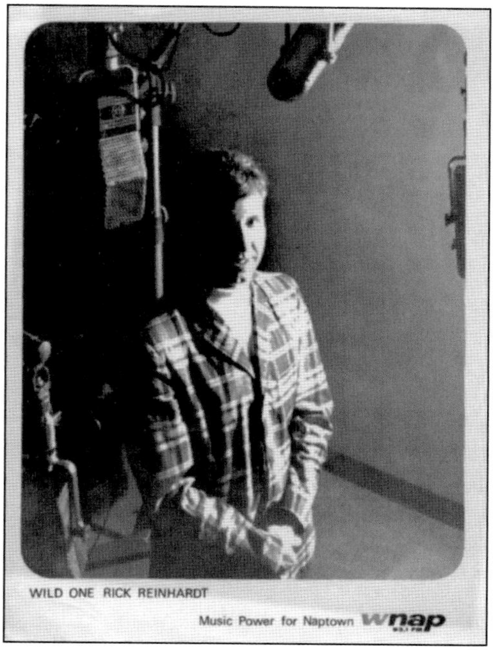

The studio was kind of austere when we first started out. It was pretty bare bones and whatever they could throw together that the engineers weren't using on WIBC. —Cris Conner

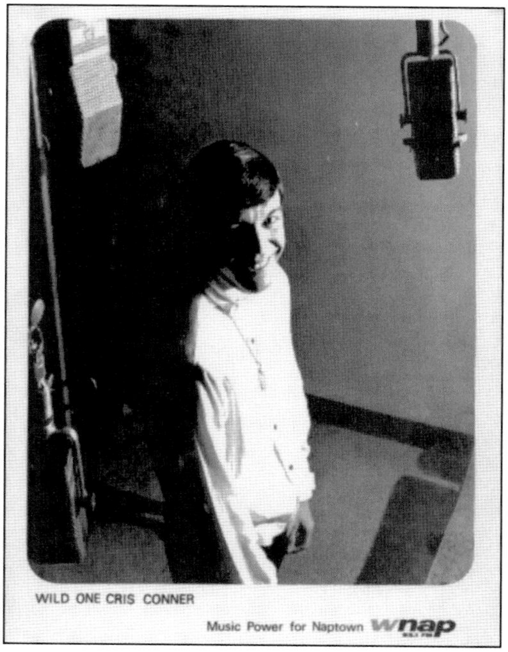

I believe very strongly in personalities. And we loaded up WNAP with the best personalities we could possibly do, but they have to match the audience. —Jim Hilliard

TV SCENE

Revamped FM Radio Outlet Promises A 'Young Sound'

By JULIA INMAN

"WE WON'T be 35 acting like 20-year-olds," said Al Stone, 24-year-old co-program director for new WNAP.

The completely revamped station, formerly WIBC-FM, went on the air yesterday morning with new call letters, a new format of hard rock music and a staff all under the age of 25.

"We're definitely teen-oriented up to the mid-20s," continued Stone, adding that the station will pick the cream of the top 40 on the music charts which appeals to the young "in" crowd.

The staff of five ranging from 20 to 24 in age has adopted "music power" as its slogan and calls itself the "Wild Ones."

They've also adopted the Indianapolis nickname "Naptown" and hope to give it a new and "good" twist.

With the exception of Stone who has a background as a professional musician and ham radio operator, the staff was recruited from Bloomington station WIUS, by WIBC and WNAP general manager William Wheatley.

On the technical side the new FM outlet has a vertically and horizontally polarized antenna for better car radio reception. The station also has plans to go into stereo in the near future.

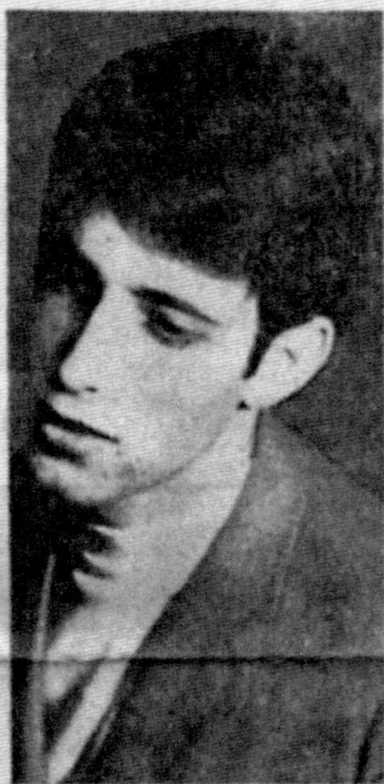

AL STONE

As for news, Stone said the outlet is the first FM station in the country to pick up the American Broadcasting Company's Contemporary Service, one of several ABC networks specially tailored to differing station formats. The service has three-minute features and some movie reviews as well as hard news.

"WE HOPE to have an innovative, vibrant, lucid type of programming and good clean humor," Stone declared.

Following rhythm and blues station WTLC-FM, WNAP is joining the trend toward specialization in local FM radio that is giving new life to FM here.

In focusing on hard rock, the revamped outlet obviously is out to challenge the top rating position of WIFE which since June also has been branching out into other types of music.

★

THE FIRST CBS Playhouse of the new season will be "The People Next Door," a 90-minute drama by J. P. Miller who wrote "Days of Wine and Roses." The play will be produced by Herbert Brodkin of Defenders fame and will be aired Oct. 15 at 9:30 p.m.

The original play will dwell on the effects on a middle-class family of the discovery that their daughter is addicted to narcotics. Lloyd Bridges, Kim Hunter, Fritz Weaver Phyllis Newman and Nehemiah Persoff head the cast.

★

THE EQUAL time suspension measure, now stalled in the House Commerce Committee, was given a shove yesterday by NBC president Julian Goodman in a wire to Representative Harley O. Staggers (D-W.Va.).

Goodman said a 1968 suspension "would enable broadcasters to present the candidates in a variety of forms and confrontations in which they wished to appear, which would include but not be limited to debates."

This newspaper coverage announces the debut of WNAP.

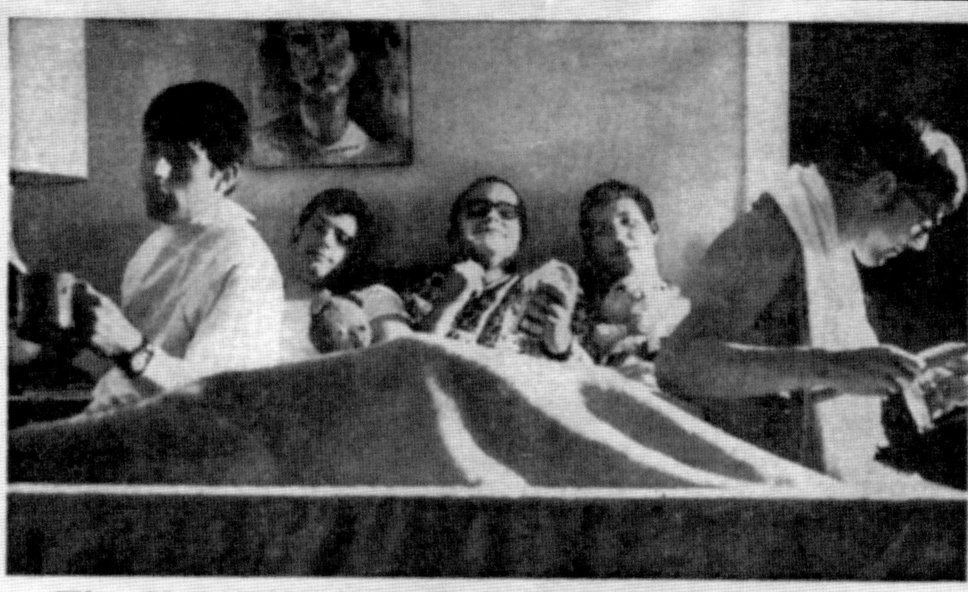

The Hard Way

The disc jockeys of new radio station WNAP-FM demonstrate their usual manner of relaxing before a hard day on the air. Al Stone, Jasen, John Gillus, Rick Reinhardt and Cris Conner comprise the teen-oriented staff of the "hard rock" station which operates on 93.1, the old WIBC-FM band.

Here is an early press clipping announcing the "new" radio station, WNAP-FM. "We won't be 35 acting like 20 year olds" was a specific dig at WIFE in the press. The first shot of the war had been fired, and it was not across the bow—it was a broadside.

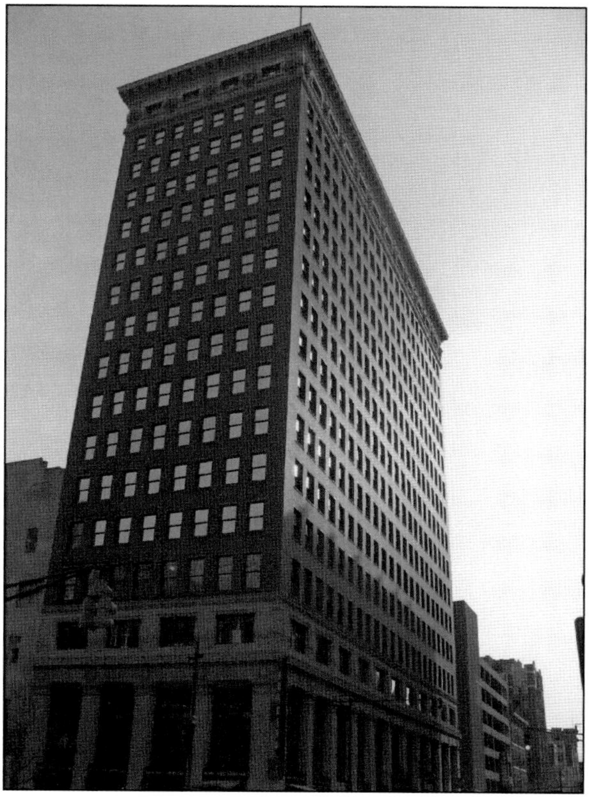

There were three logical reasons for the WNAP transmitter and tower to be located atop the Merchants Bank Building at the intersection of Washington and Meridian Streets: It was centrally located, it was one of the highest points in Indianapolis, and Richard M. Fairbanks was part owner of the bank, thus there was no red tape.

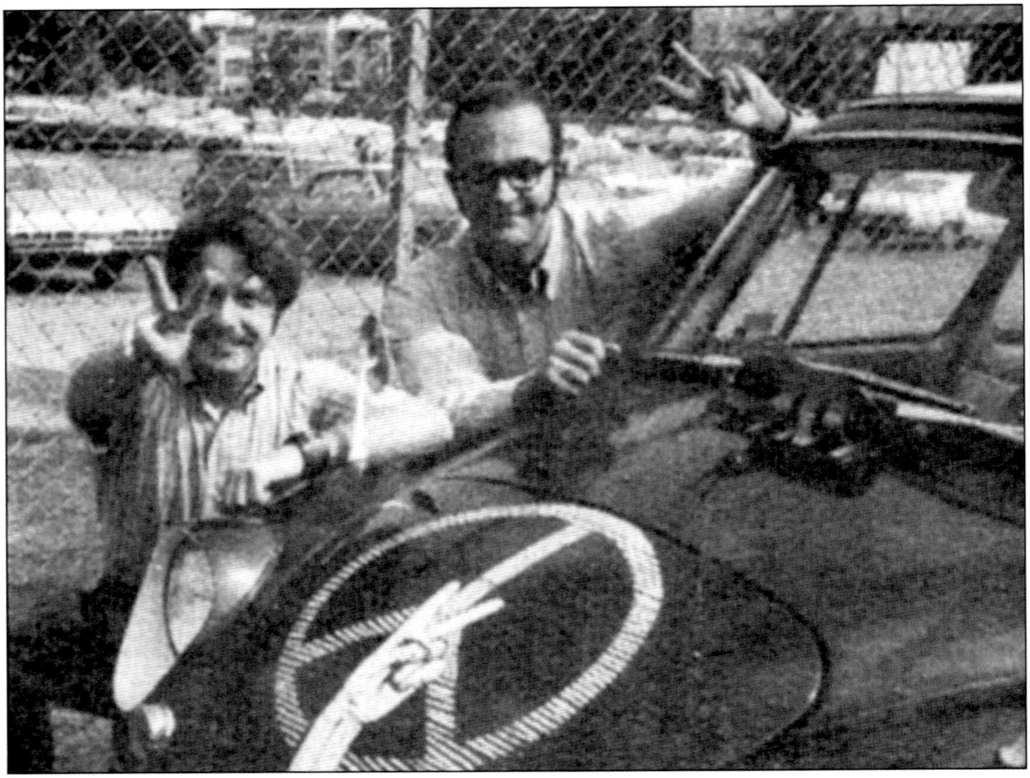

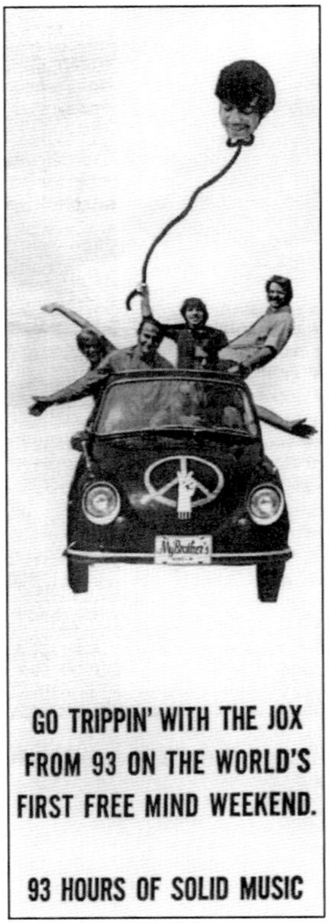

The "Grossmobile" was an Isetta convertible painted to promote the station around town. The helium balloon head is, of course, that of Cris Conner.

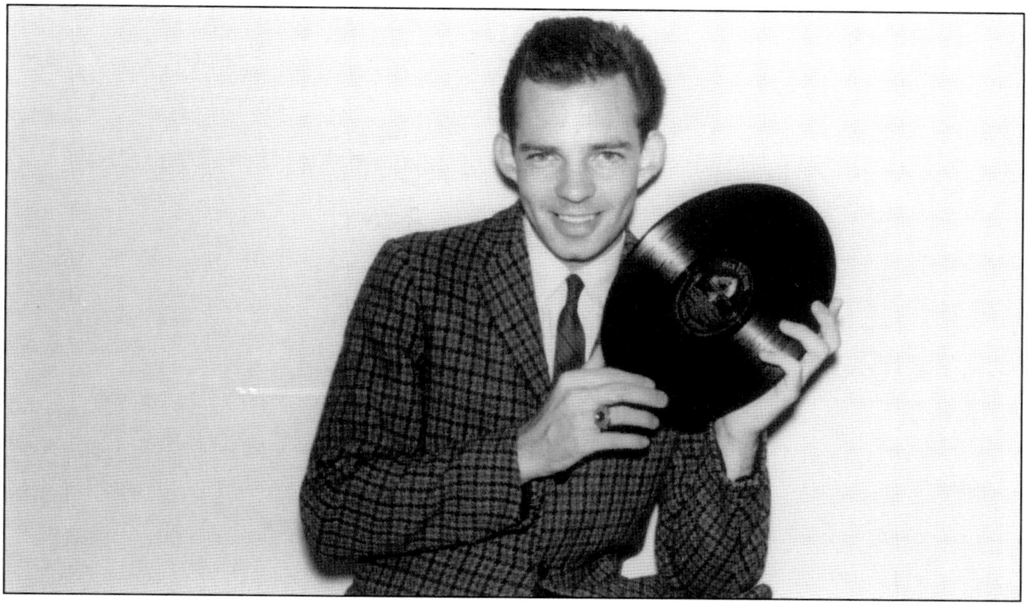

Mike "Griff" Griffin was also originally slated to be among the initial on-air staff of WNAP. Griffin and Cris Conner had worked together at WJVA, a station in South Bend, Indiana. Despite saving up leave time to get out early, Griffin's commitment to Naval Air Reserve tied him up until two weeks after WNAP went on the air. Even though he missed the launch, he would become one of WNAP's best-known deejays.

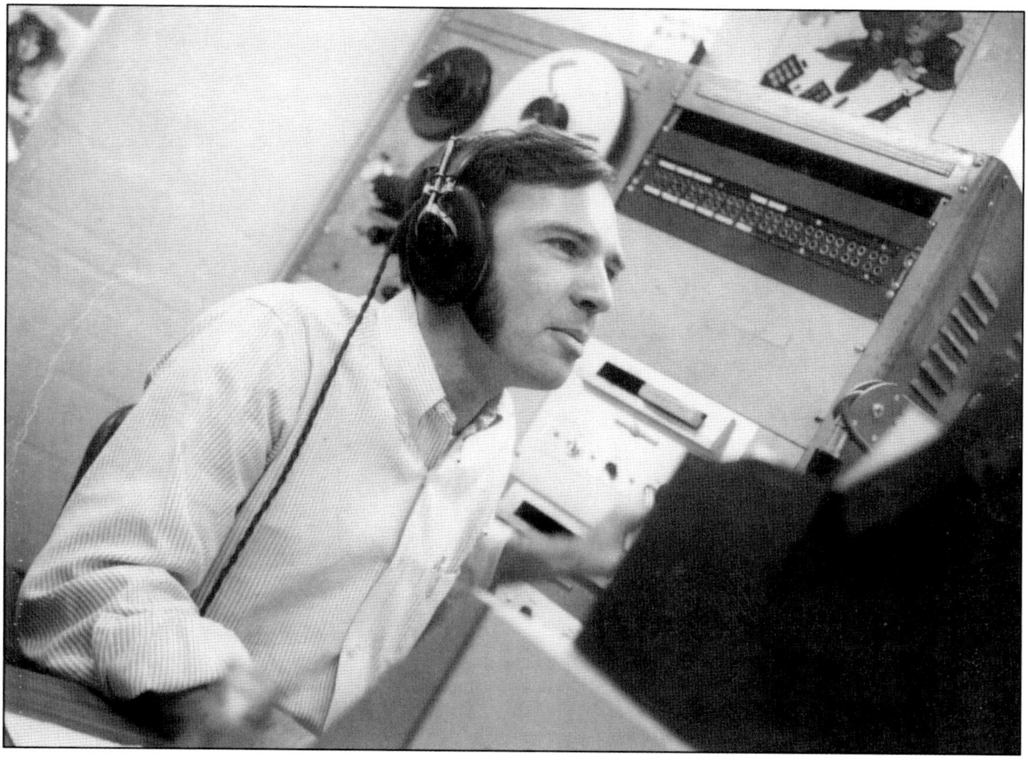

THE YOUTH RACE

The best indicator of the new strength of FM is the head-on collision for the teens and young-adult audience by WIFE, WNAP-FM and WTLC-FM.

The latest rating information was taken last April and May. ARB and Pulse contradict one another in some respects, a not uncommon occurrence in the highly unscientific business of ratings, but even so, both rating companies show WNAP and WTLC breathing hard down WIFE's neck, and with the FMs dominating the teen-age audience in certain segments of the day.

The ARB rating went so far as to indicate WNAP as the undisputed leader in total evening audience among all stations for the 27-county area surrounding Indianapolis and considerably ahead of WIFE in the metropolitan area — eight counties.

But this isn't intended as any kind of racing form to arm one broadcast space salesman against another.

It is intended to point up the fact that Major Armstrong's baby — stillborn — has been resuscitated and is growing into a giant.

With broadcast revenues down in general, the arrival of FM on the scene has put some of the older stations in a bind.

WXLW, for example, which has had its troubles in recent years, is at a low ebb in audience. The owner, Robert Enoch, denies one rumor that the station may close. He said there will be changes to announce soon. Among the rumored changes is the possible sale to one of three companies which are reported to be interested.

In the same vein, WGEE, which had gone along for years on a hybrid format of country and western part of the day and rhythm and blues the rest, also is having its problems.

WIRE swiped WGEE's country and western audience and WTLC-FM grabbed the rhythm-blues fans.

A late-1960s blurb in the paper commented about the shift in ratings in the Indianapolis market. WIFE had been without serious competition for several years, and within six months of each other, WTLC and WNAP had cropped up like nasty weeds.

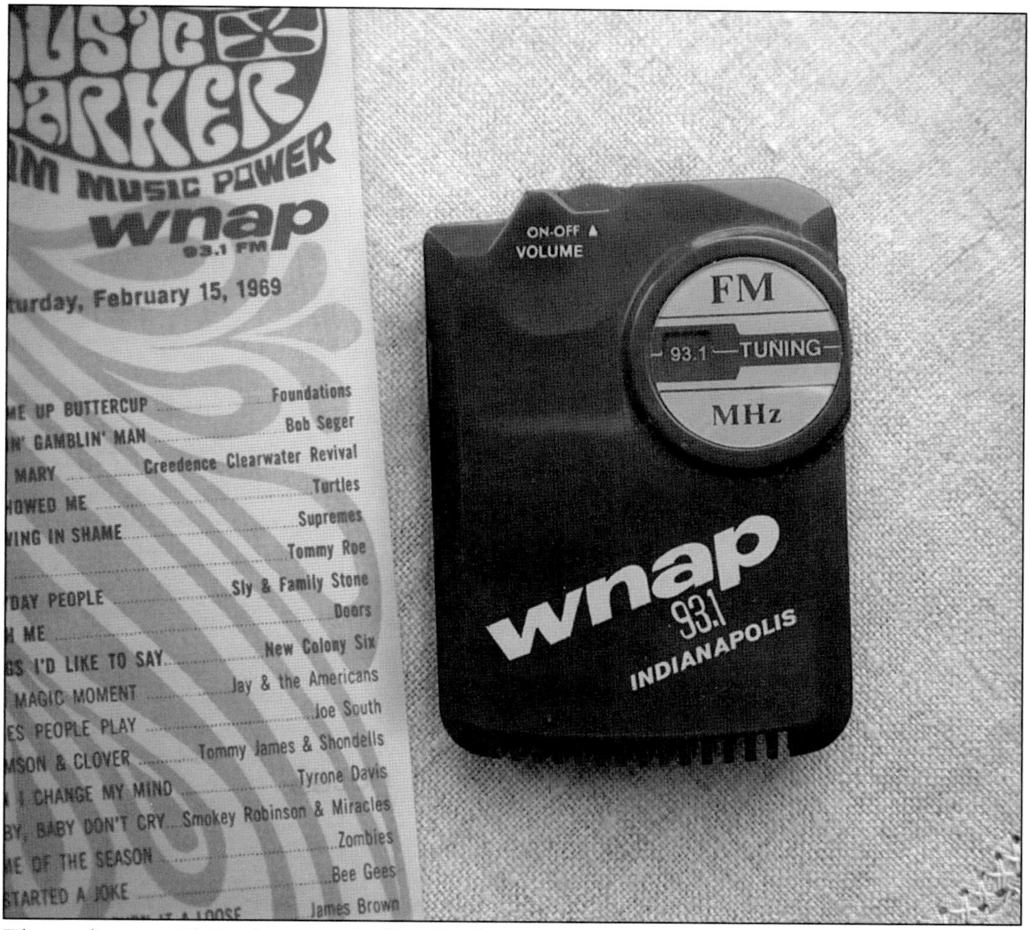

The push to put FM radios into the hands of consumers began. (Composite photo.)

FM radio was something that people might listen to in their home on a sort of a console, or a special radio. But most of the radios that people of my generation were listening to were either in automobiles, or they were transistor radios that they had in their bedrooms, or they would take with them for outside activity. And in those days, all it was was AM. —Tom Cochrun

At WNAP, one of our first promotions was really unique. We decided to give away a bunch of little, cheap FM radios that could only get 93.1. —Al Stone

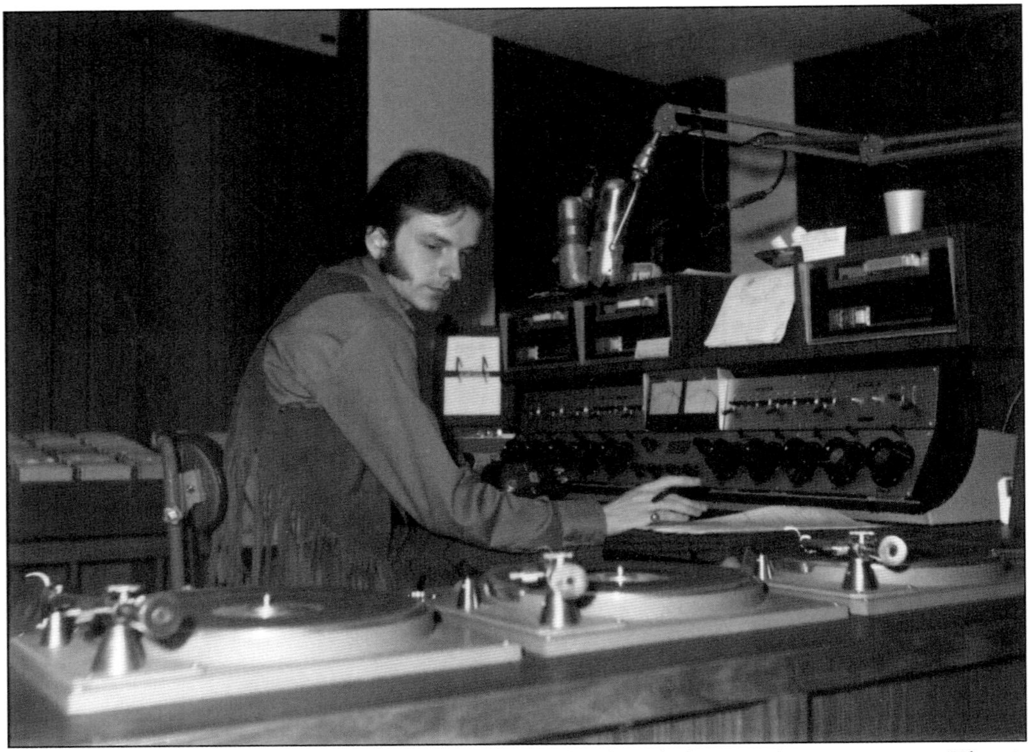

On New Year's Day in 1970, WNAP ushered in the decade by switching to stereo. This is Steve King, who was on the air middays, with a pair of RCA 77DX microphones to take advantage of the stereo separation. Note the three new cart machines adorning the space over the mixing board.

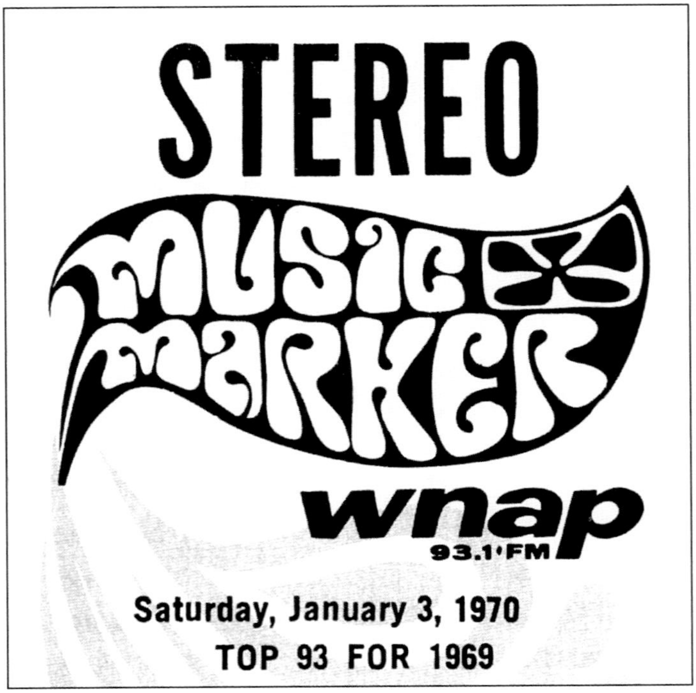

Well, for us, being inside the studio, going stereo was "oh, man, this is great music! It's coming out of both speakers." —Cris Conner

And we saved up for bigger speaker monitors so we could make it really loud. And when it got to the point that they could hear it all over the building, they screamed and yelled about smaller amplifiers and turning it down. —Mike Griffin

WNAP still did not have many commercials to run in 1970, so having 93 hours without ads was not a problem. Sponsors, though, became insulted with the original "commercial free" boast, so it was changed to "non-stop music" to make everyone happy.

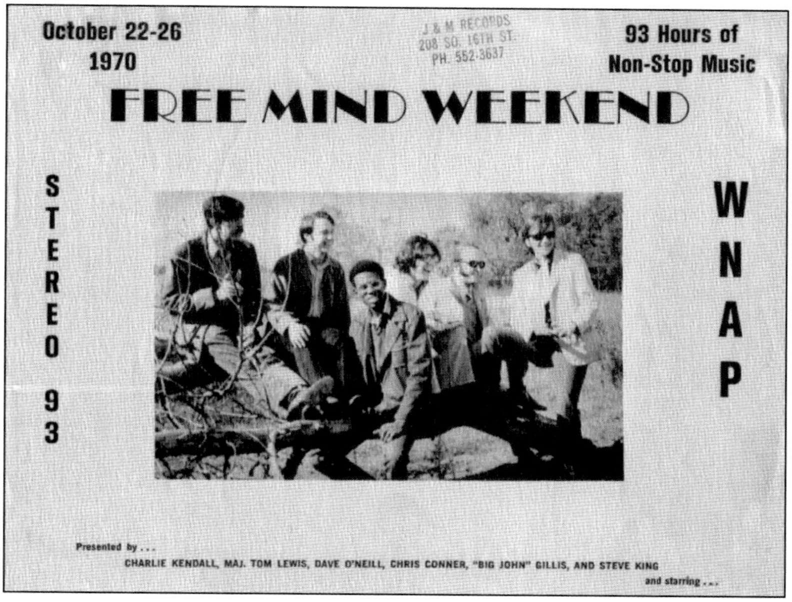

The "Free Mind Weekends" typical for NAP, occasionally we'd take a whole weekend and play nothing but obscure album cuts and long songs that you'd never hear on AM. —Al Stone

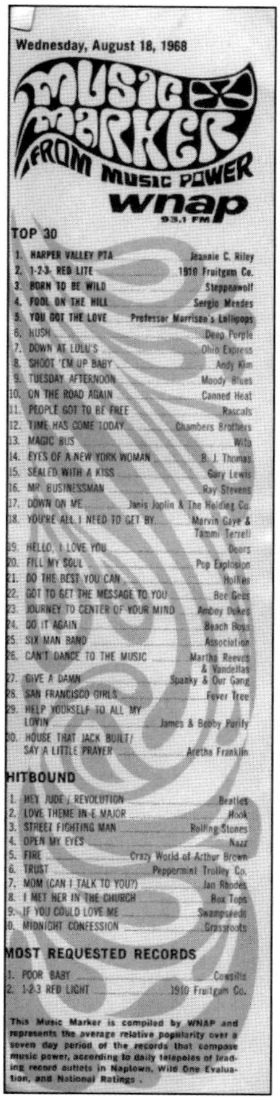

Wednesday, August 18, 1968

MUSIC MARKER FROM MUSIC POWER wnap 93.1 FM

TOP 30

1. HARPER VALLEY PTA — Jeannie C. Riley
2. 1-2-3- RED LITE — 1910 Fruitgum Co.
3. BORN TO BE WILD — Steppenwolf
4. FOOL ON THE HILL — Sergio Mendes
5. YOU GOT THE LOVE — Professor Morrison's Lollipops
6. HUSH — Deep Purple
7. DOWN AT LULU'S — Ohio Express
8. SHOOT 'EM UP BABY — Andy Kim
9. TUESDAY AFTERNOON — Moody Blues
10. ON THE ROAD AGAIN — Canned Heat
11. PEOPLE GOT TO BE FREE — Rascals
12. TIME HAS COME TODAY — Chambers Brothers
13. MAGIC BUS — Who
14. EYES OF A NEW YORK WOMAN — B. J. Thomas
15. SEALED WITH A KISS — Gary Lewis
16. MR. BUSINESSMAN — Ray Stevens
17. DOWN ON ME — Janis Joplin & The Holding Co.
18. YOU'RE ALL I NEED TO GET BY — Marvin Gaye & Tammi Terrell
19. HELLO, I LOVE YOU — Doors
20. FILL MY SOUL — Pop Explosion
21. DO THE BEST YOU CAN — Hollies
22. GOT TO GET THE MESSAGE TO YOU — Bee Gees
23. JOURNEY TO CENTER OF YOUR MIND — Amboy Dukes
24. DO IT AGAIN — Beach Boys
25. SIX MAN BAND — Association
26. CAN'T DANCE TO THE MUSIC — Martha Reeves & Vandellas
27. GIVE A DAMN — Spanky & Our Gang
28. SAN FRANCISCO GIRLS — Fever Tree
29. HELP YOURSELF TO ALL MY LOVIN — James & Bobby Purify
30. HOUSE THAT JACK BUILT/ SAY A LITTLE PRAYER — Aretha Franklin

HITBOUND

1. HEY JUDE / REVOLUTION — Beatles
2. LOVE THEME IN E MAJOR — Hook
3. STREET FIGHTING MAN — Rolling Stones
4. OPEN MY EYES — Nazz
5. FIRE — Crazy World of Arthur Brown
6. TRUST — Peppermint Trolley Co.
7. MOM (CAN I TALK TO YOU?) — Jan Rhodes
8. I MET HER IN THE CHURCH — Box Tops
9. IF YOU COULD LOVE ME — Swampseeds
10. MIDNIGHT CONFESSION — Grassroots

MOST REQUESTED RECORDS

1. POOR BABY — Cowsills
2. 1-2-3 RED LIGHT — 1910 Fruitgum Co.

This Music Marker is compiled by WNAP and represents the average relative popularity over a seven day period of the records that compose music power, according to daily telepoles of leading record outlets in Naptown, Wild One Evaluation, and National Ratings.

While WNAP would gain the reputation for daring and obscure playlists, the dirty little secret is that when it first went on the air, the station played pretty much the same bubble-gum music as WIFE.

RADIO

FM Stations

*Stereo

88.7—WICR-FM

90.1—WIAN-FM

91.3—WHJE1FM
(Daytime only)

92.3—WTTV-FM

93.1—WNAP-FM
(6 a.m.—2 a.m.)

94.7—WFBM-FM
(24 hours)

*95.5—WFMS-FM
(6 a.m.—1 a.m.)

*95.9—WIFN-FM

*97.1—WSVL-FM

*98.3—WJMK-FM
(6 a.m.—11 p.m.)

99.5—WSMJ-FM
(6 a.m.—2 a.m.)

100.9—WNON-FM
(6 a.m.-11 P.M.)

103.3—WGEE-FM
(6 a.m.—Midnight)

103.7—WFIU-FM

104.5—WAJC-FM
(4 p.m.—Midnight)

105.7—WTLC-FM
(11 a.m.—Midnight)

*107.9—WIFE-FM
(24 hours)

AM Stations

810—WATI
(7 a.m.—6 p.m.)

950—WXLW
(6 a.m.—6 p.m.)

1070—WIBC
(24 hours)

1260—WFBM
(24 hours)

1310—WIFE
(24 hours)

1430—WIRE
(24 hours)

1500—WBRI
(7 a.m.—6 p.m.)

1590—WGEE
(daylight hours)

Radio, TV Sports Menu

TODAY

8:00—WNON-FM Baseball: Cincinnati Reds vs. St. Louis Cardinals.

The battle between WIFE-AM and WNAP-FM had begun.

What they did was they counter-programmed in the late Sixties against WIFE-AM at 1310 with WNAP on their FM. Their signal power was far superior, clarity of sound far superior, high fidelity, which we didn't have with WIFE on the AM. They created a strong array of disc jockeys just like we had at WIFE. And what they did was, you know what, they sat back and they listened. And they knew they were getting their brains beat out, but they were smart, and they paid attention, and they counter-programmed, and they came right at us. —Jack Marsella

There was an evolution in the creation of the WNAP Buzzard logo.

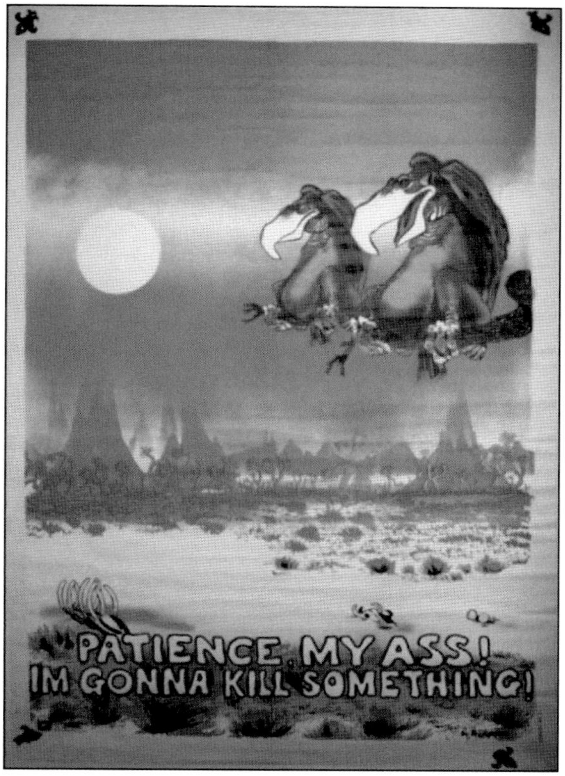

The Buzzard came about because we didn't really have an image yet that we had defined at the radio station, something that you could really hang your hat on that was promotable. And there was a poster around at the time of two vultures sitting on a limb, one saying to the other, "Patience, my ass, I'm gonna kill something." —Cris Conner

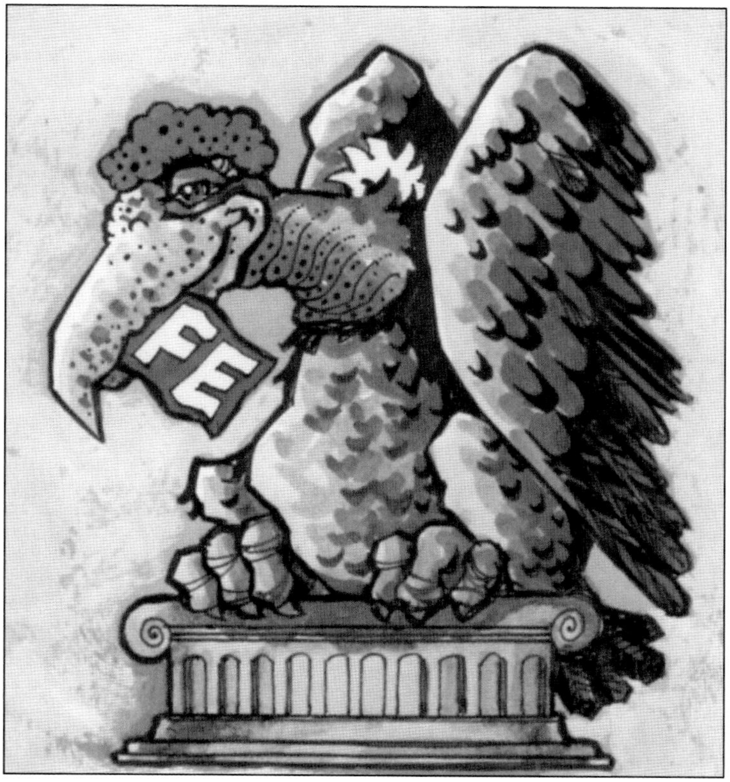

And we thought that's perfect—that's what we were about. Our mission was to knock off WIFE. And it's like, let's take it to the next level. Let's make that our mascot. Let's make that our representative. That's the most repulsive—you know, the establishment will hate that image. They will absolutely hate it. —Mike Griffin

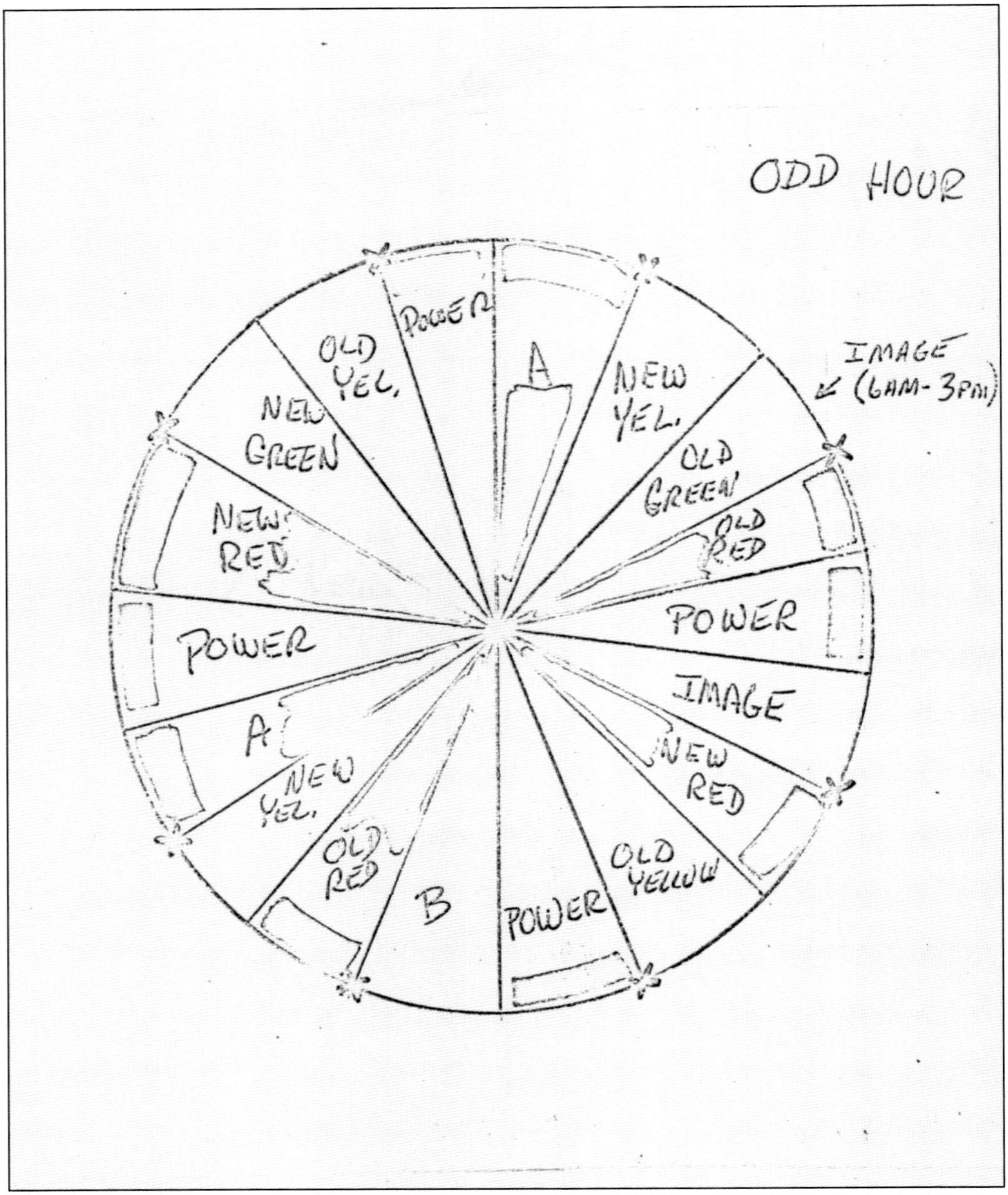

It is interesting to compare the "programming clocks" of WIFE and WNAP. A clock, like the WIFE-AM one seen here, is a graphical representation of what the deejay is supposed to be playing at a given point in time.

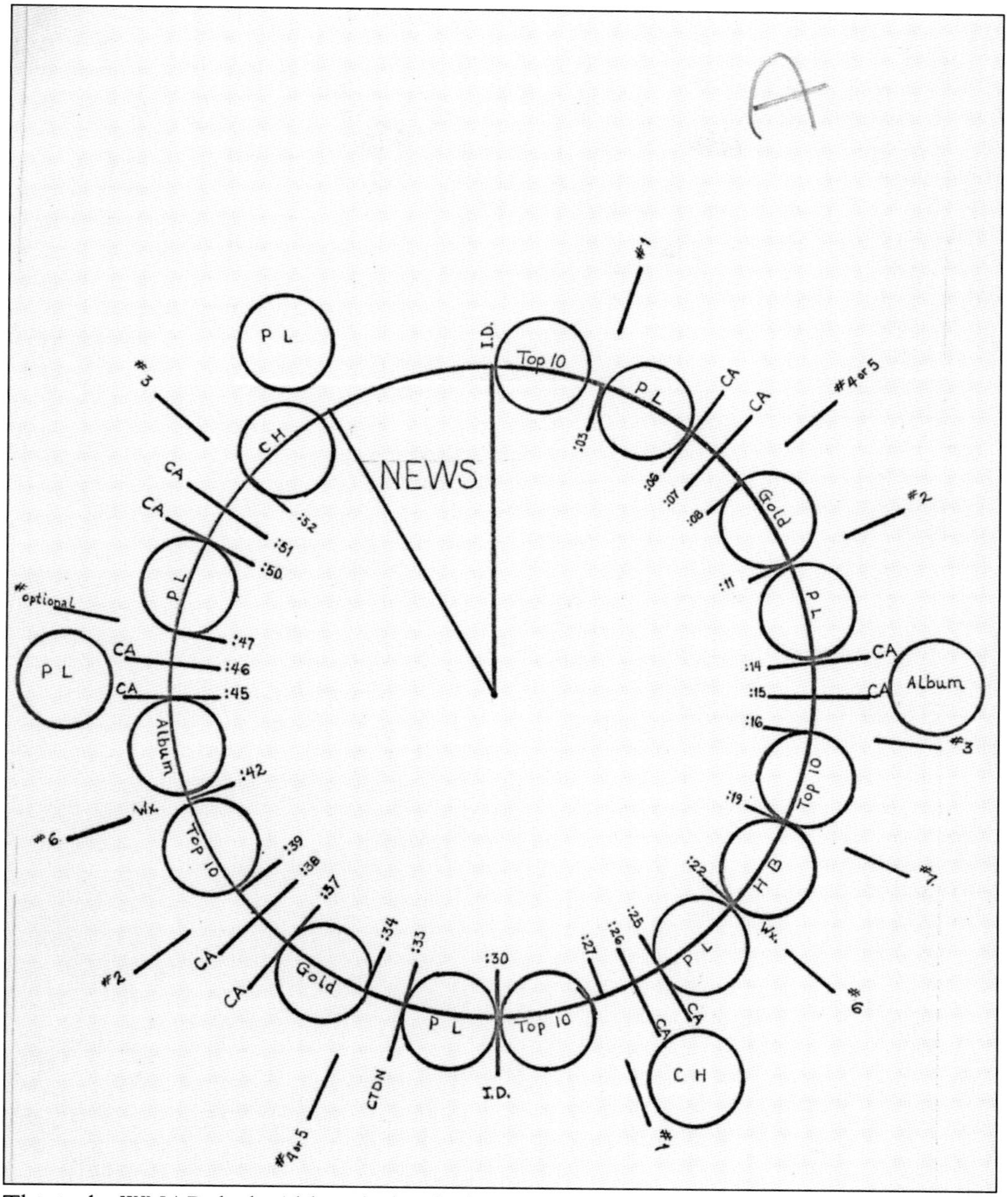

This is the WNAP clock. Although the clocks on the previous page and here are not that much different, notice at 3 and 9 on the dial are the inclusion of album cuts. There is great debate about how many album cuts WNAP actually played per hour. The collective memory of the listening audience is that WNAP was a very radical station that constantly played albums. While the concentration of album cuts would vary during the course of the station's life, by and large they were really only playing about two per hour. In 1973, WNAP pretty much removed album cuts from the playlist by order of George Johns, who said, "there's no money in deep cuts."

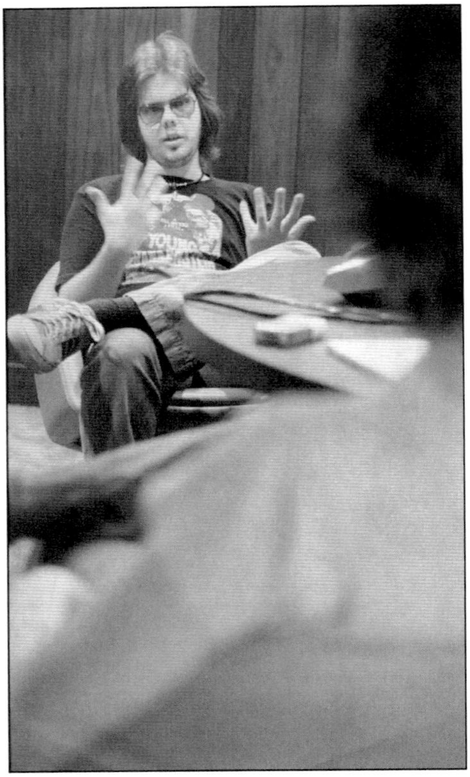

In 1972, Michael D. Hanks came to WNAP with the on-air name of "Buster Bodine." Hanks's brother was Charles "Chuck Riley" Hanks, the afternoon-drive deejay at sister station WIBC. The two of them would put together the most famous station ID in town, "The wrath of the buzzard, WNAP, Indianapolis," with Chuck providing the growly voice-over and Buster cutting it together with the sound of thunderbolts.

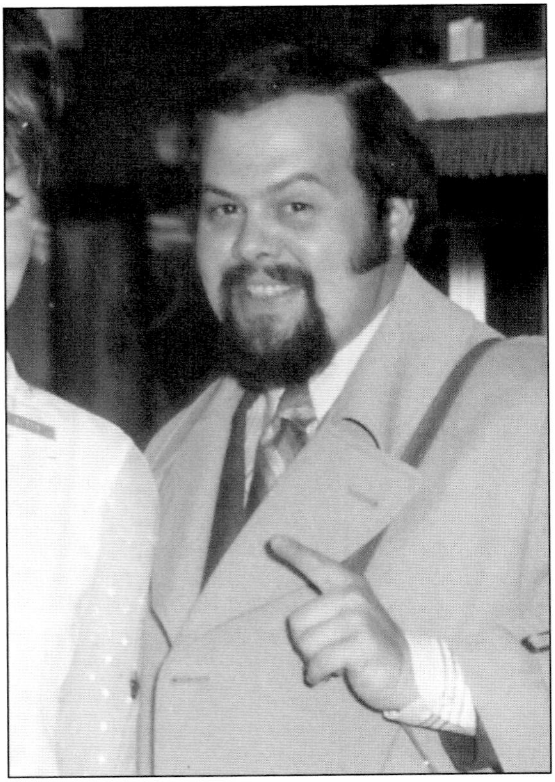

Chuck Riley was the station's bon vivant. He had afternoon drive on WIBC with a show called *The Life of Riley*. Because the station wanted him to sound very grounded and connected to Indianapolis, his on-air name paid homage to native son James Whitcomb Riley.

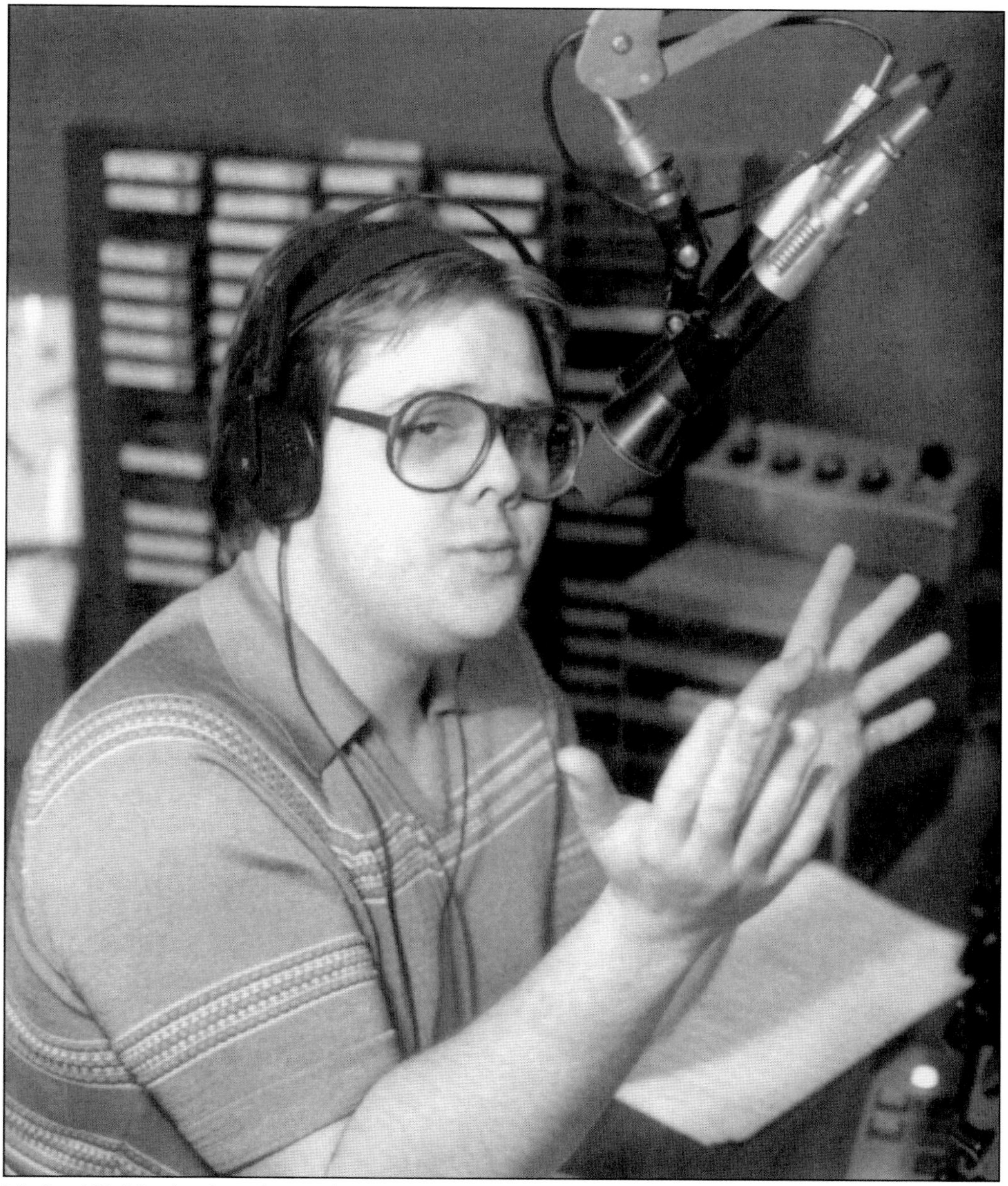

Michael D. Hanks thought up the name Buster Bodine during a cross-country drive. He wanted to do something with the name Jethro Bodine, a character from the TV show "The Beverly Hillbillies." A moment later, he drove by a Buster Brown shoe billboard, and it came to him.

Buster Bodine was a hippie truck driver. I envisioned him as some blue-collar guy that wasn't that bright, yet was bright enough to know certain things to make people think he was a little brighter than he was. And I mean there was a whole thing constantly going on in my head about the character. And that's all it was to me was just a character. It has nothing to do with me. And it's the furthest—I mean, I'm a conservative family guy, and children, the whole thing. But when I would go on the air and do this character, it's like zipping up a suit that said "Buster Bodine" on it. I turned into that guy. —Michael D. Hanks

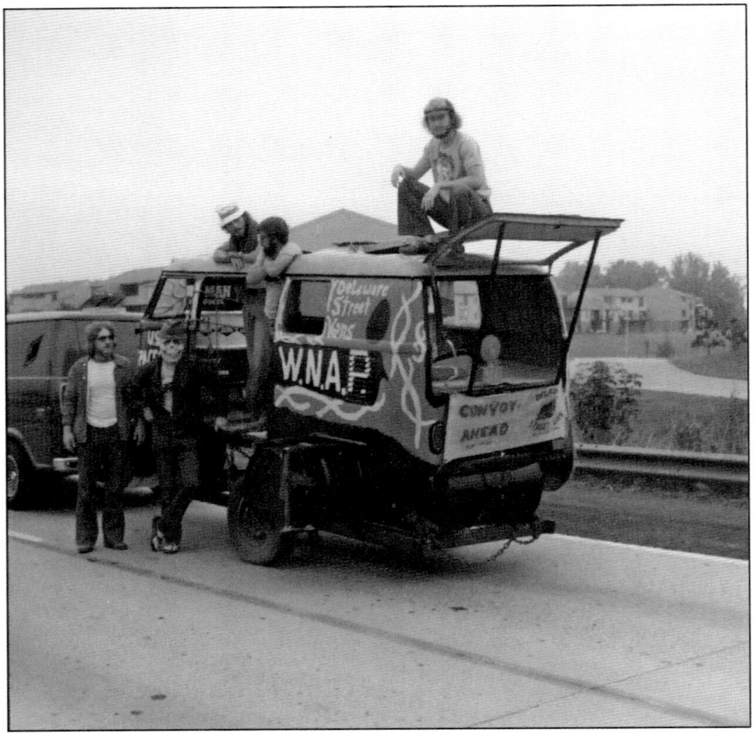

The WNAP Raft Race was a fixture in Broad Ripple from 1973 to 1982. The rafts would start on the White River near Landings Apartments near Keystone Avenue and wind around the river bends to a finish line at Broad Ripple Park, where live bands performed. Special consideration was given to rafts that incorporated either WNAP or the buzzard into the theme.

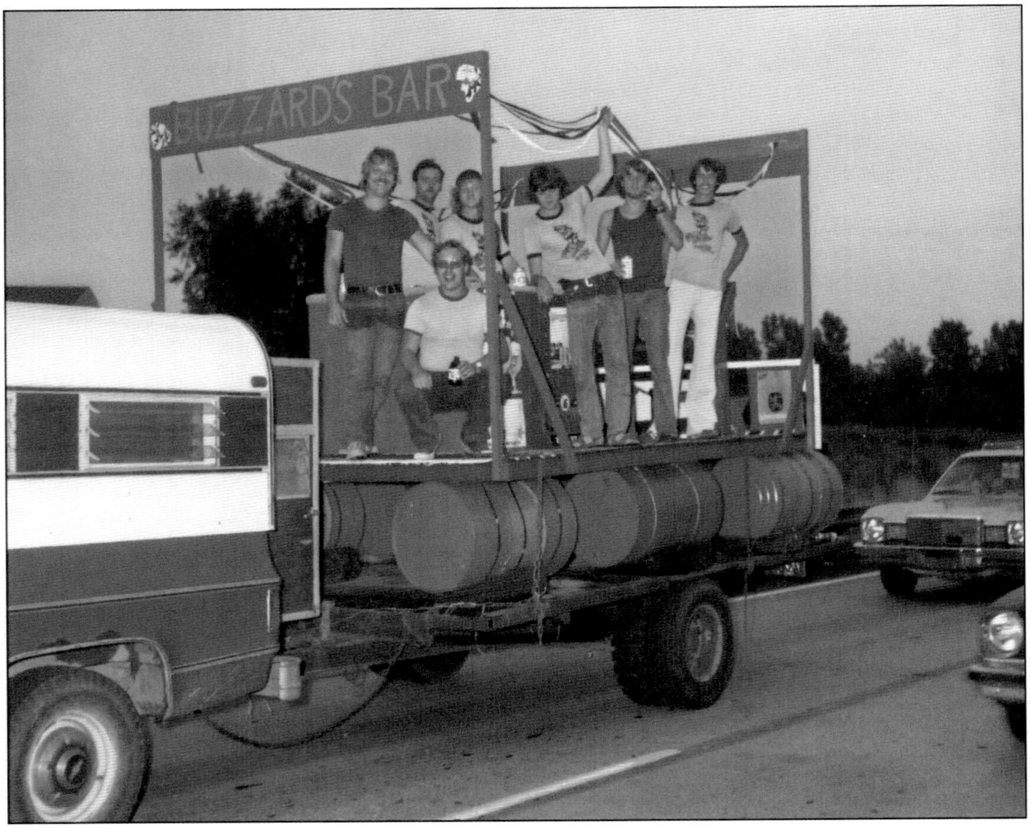

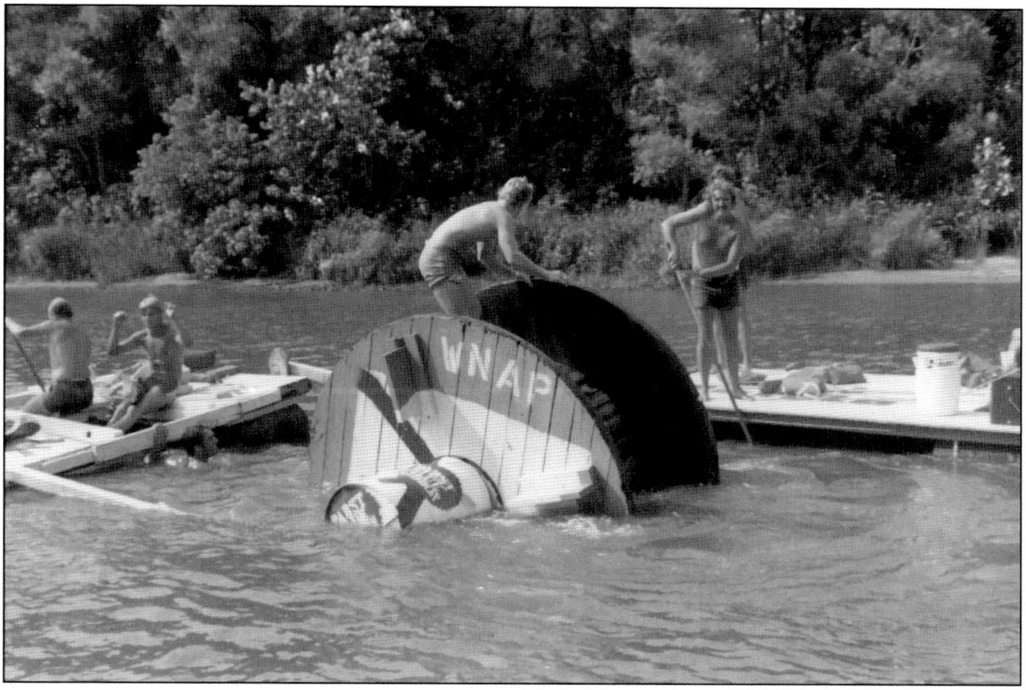

Not all rafts that entered were either safe or practical, but that was irrelevant. The race also had special guests from time to time. Jocko Conley is seen at right enjoying time with a Hawaiian Tropic girl.

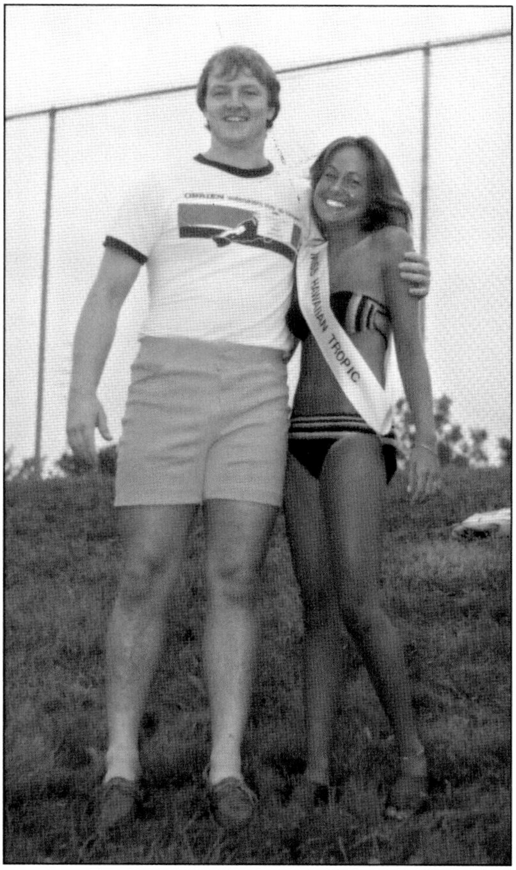

We had the WNAP Raft Race. The greatest promotion ever created. That wasn't a contest. You didn't win anything. I think we made a prize. What was the prize? I think we gave something for the best floaters. It didn't make any difference. It's a promotion, it's a life event. —Jim Hilliard

The Raft Race fit into the strategy because it got on TV. If you have a radio station, and you're on TV, and you're on the news, and, you know, TV time's expensive, and we're spending our money on other things. —George Johns

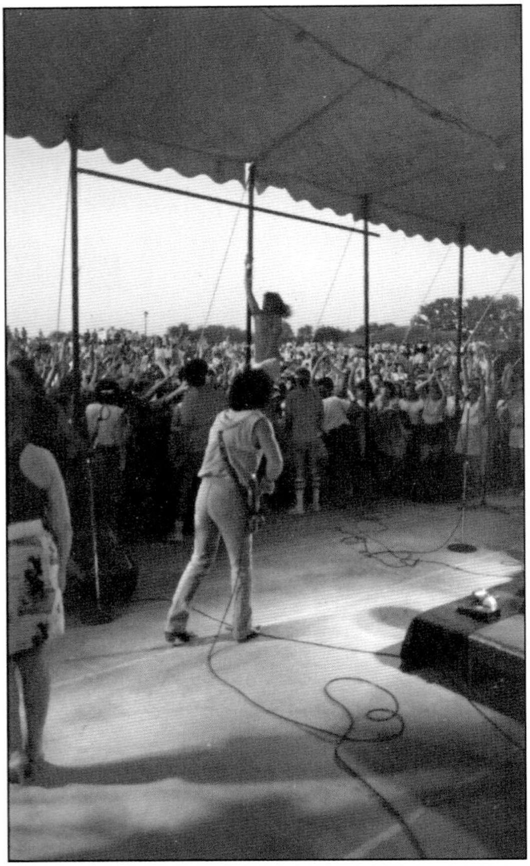

By its second year, the WNAP Raft Race had already become wildly successful, drawing estimated crowds as high as 100,000 to little Broad Ripple Park. Local musician Henry Lee Summer is shown at left performing at the stage in Broad Ripple Park.

When we started out our first year, we had 150 rafts. We didn't know what we were going to do. The second year we did it, we had over 400 rafts. And the concert was the deal. We partied all day down the river, and then you got out at the park and you had an all-day concert going on. —Cris Conner

The Broad Ripple village people, they were not happy. They had drunks in their yards. They'd take their lawn furniture to go to the concert. But they still let us come back every year. —Cris Conner

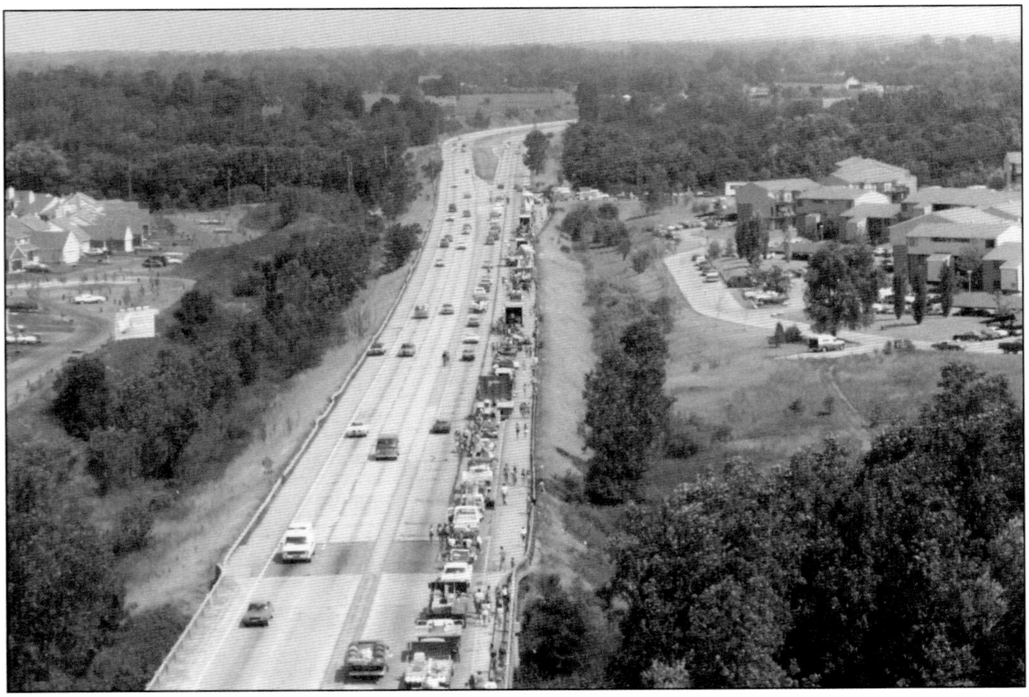

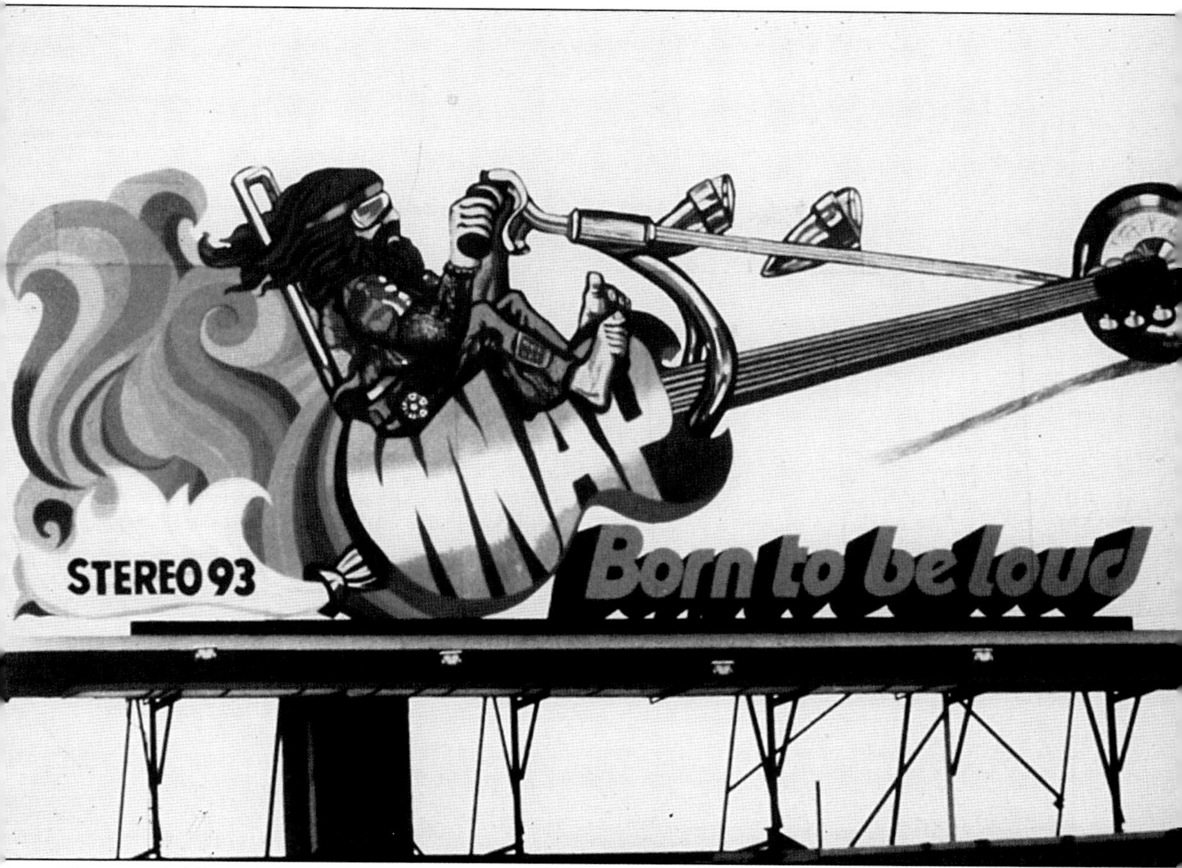

This is a billboard for WNAP from the mid-1970s. With the arrival of George Johns came a movement away from album cuts altogether with the exception of *The 25th Hour*, a midnight program that played an album from beginning to end without interruption; otherwise, the playlist tended toward rocking Top 40. The image of the station still needed to be loud and edgy, hence the theme of this billboard.

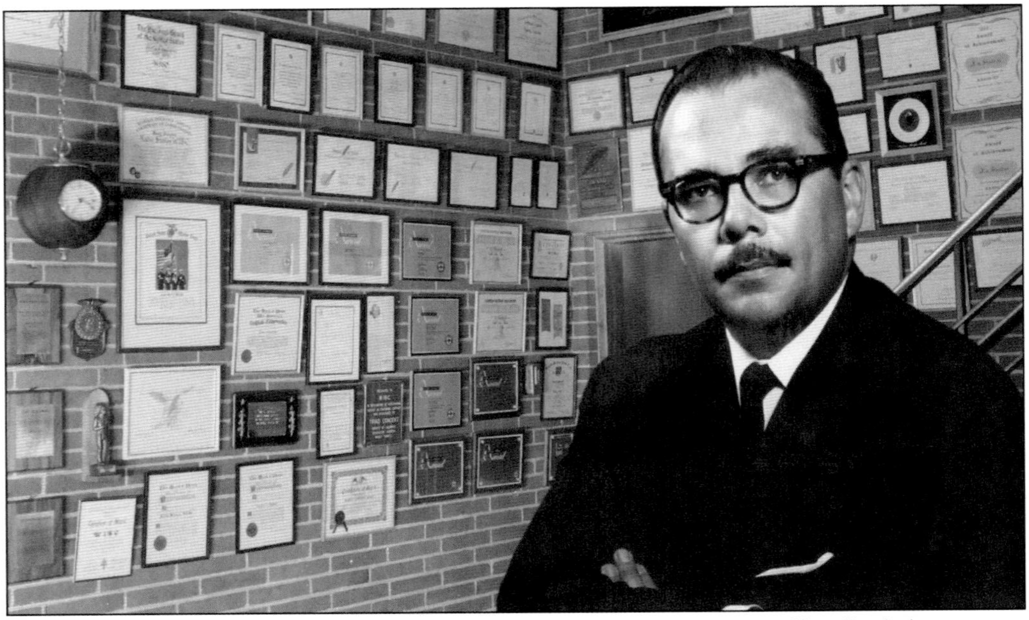

At Fairbanks Broadcasting, WIBC was known for its easy-listening music and hardboiled newscasts, and WNAP was filled with counterculture revolutionaries. The two staffs shared the same building, and often the staffs clashed. Fred Heckman, pictured, was the news director of WIBC.

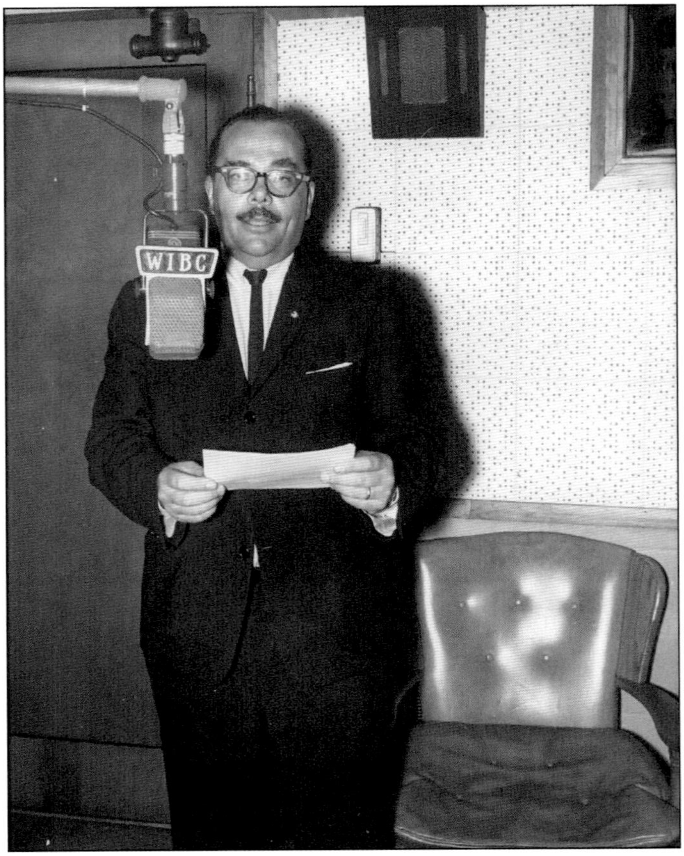

Everybody had to come in through the same door. The front of the building was Fred Heckman's property. And people who remember Fred Heckman know that he was no b.s. He was hard-core conservative. And he would rather knock you out if you came in with hair below your ears. —Cris Conner

Fred Heckman used to refer to WNAP as the "tinker toy in the back." —Ann Craig-Cinnamon

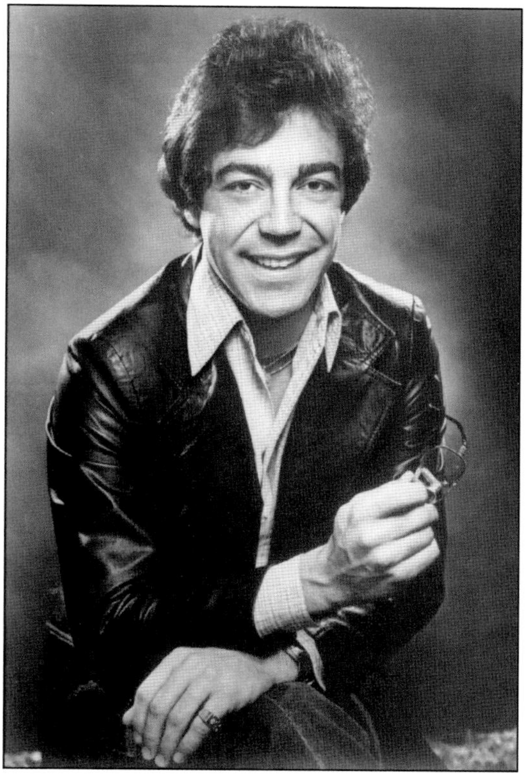

Cris Conner retired from radio in 1968 after working at WJVA in South Bend, Indiana. He moved back to South Haven, Michigan, and was starting a new career selling Cadillacs when the phone rang. Would he be interested in coming to work at a new station in Indianapolis called WNAP? He did not hesitate answering yes. Conner moved up the ladder, becoming program director and one of the top on-air personalities in the history of Indianapolis broadcasting.

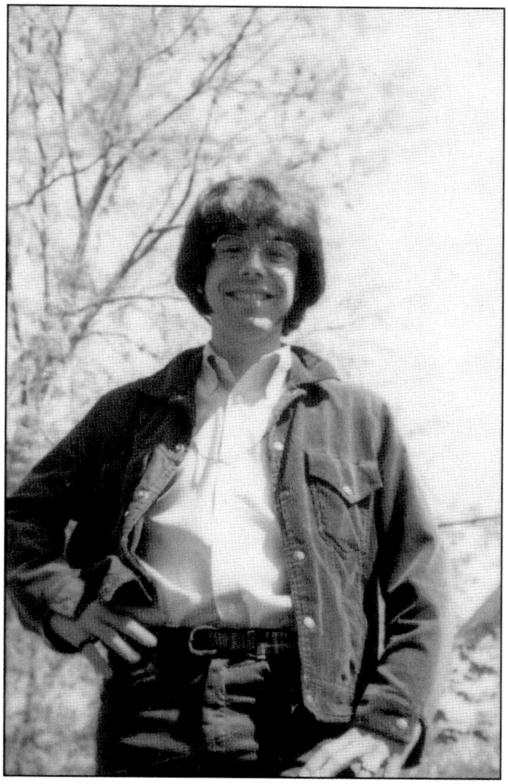

Cris Conner was the heart of that radio station. He was the guy that said, "This is hip, this is not hip. Well, this is not that bad, we're going to make it hip." He was that good. —Buster Bodine

The guy is a genius. The guy is the hippest guy to this day. He is more alive, more alert of the world, he is more tuned in, if you will, than most anybody you'll bump into. —Jim Hilliard

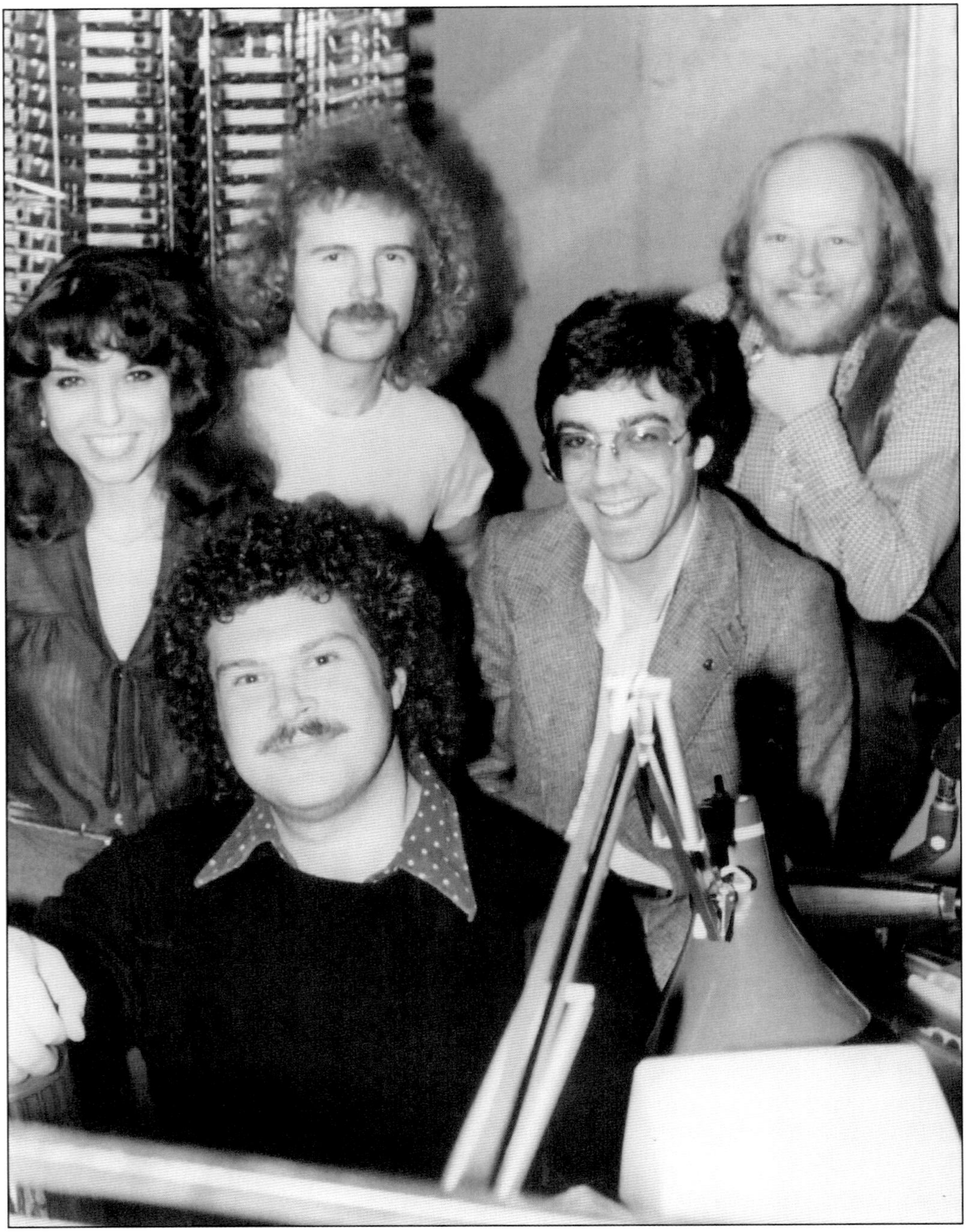

In the mid-1970s, Cris Conner moved to mornings, but he did not want to do it solo and thought it would be more interesting to surround himself with interesting personalities who would complement one another. This was the inception of what would later become known as the "zoo format," now a standard of rock radio stations. Pictured here from left to right are Ann Craig-Cinnamon, Adam Smasher (foreground), Freddie Fever, Cris Conner, and Ron Below.

Tom Cochrun joined Fairbanks Broadcasting in 1968, working for Fred Heckman as a reporter in the news department. He was quickly channeled to doing news for WNAP. Cochrun had a solid programming mind for radio and played a huge part in WNAP's information presentation of news and public-service and special programming events.

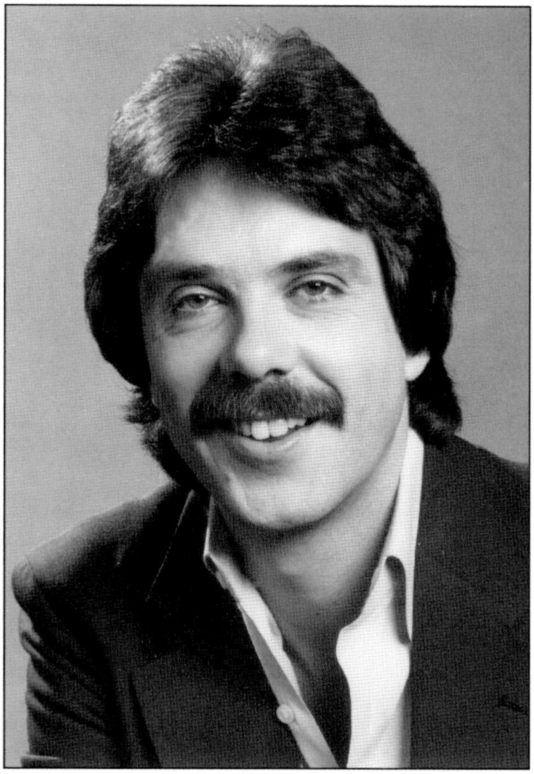

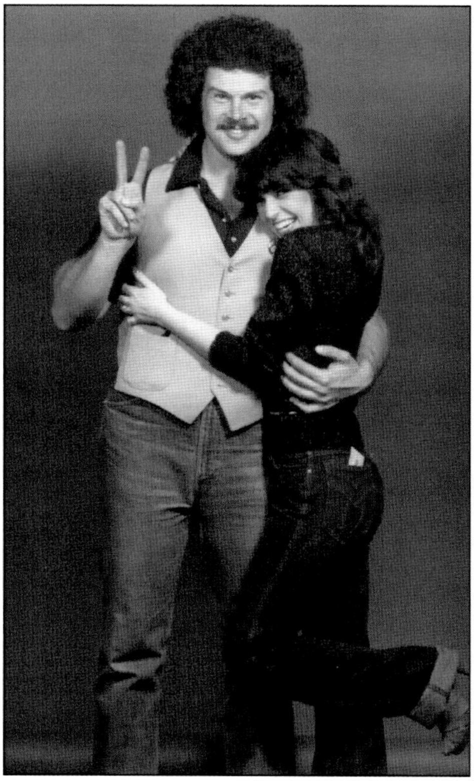

Asher "Adam Smasher" Benrubi was the lead singer of Indianapolis rock band Roadmaster and had never done radio. When he left the band, Cris Conner saw promise and offered him a job at WNAP to see what would happen. "Adam Smasher" was, well, a smash. His departure, though, was controversial when he left to go to WZPL, known then as KISS99, a disco station. The move violated the non-compete clause of his contract with WNAP. Adam is shown here with Ann Craig-Cinnamon.

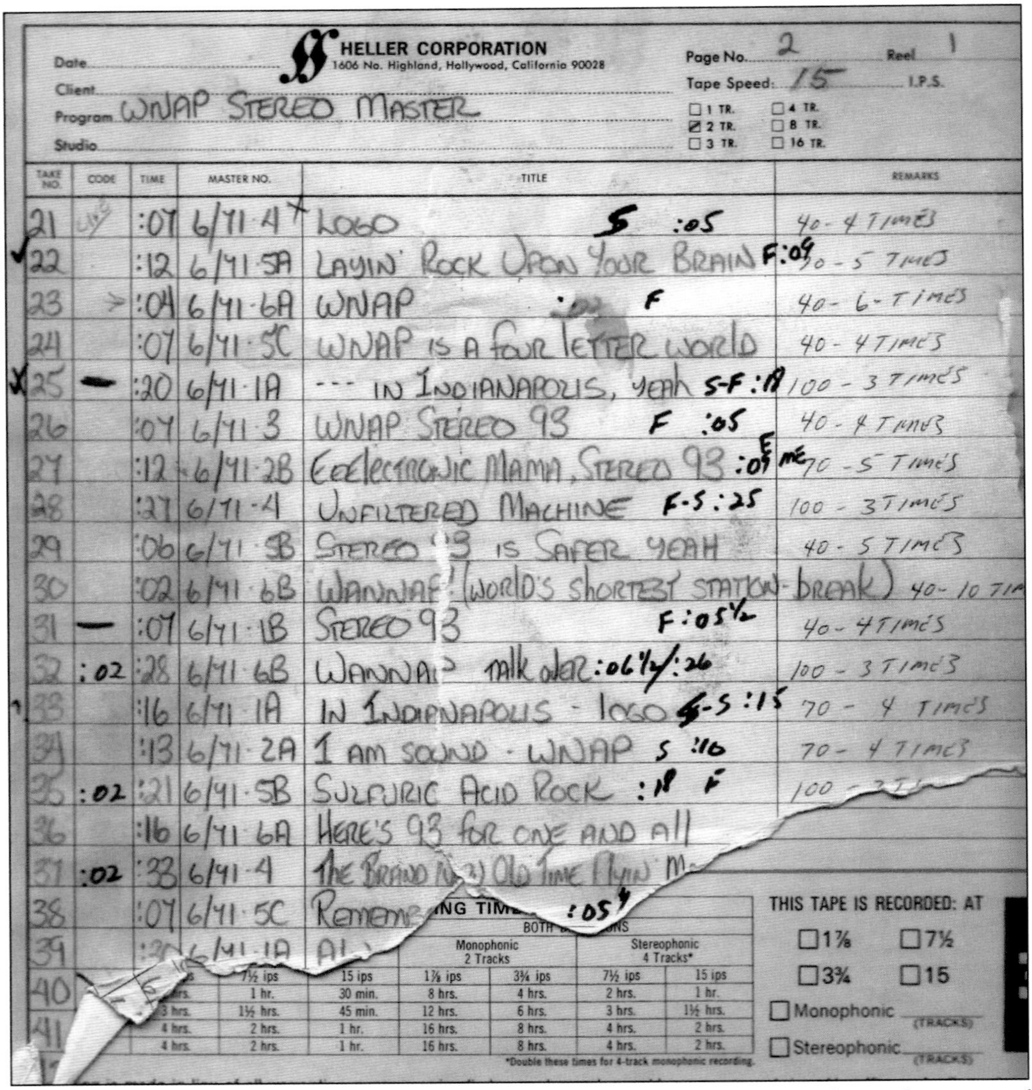

HELLER CORPORATION
1606 No. Highland, Hollywood, California 90028

Date
Client
Program WNAP STEREO MASTER
Studio

Page No. 2 Reel 1
Tape Speed: 15 I.P.S.
☐ 1 TR. ☐ 4 TR.
☒ 2 TR. ☐ 8 TR.
☐ 3 TR. ☐ 16 TR.

TAKE NO.	CODE	TIME	MASTER NO.	TITLE	REMARKS
21		:07	6/71·4	LOGO S :05	40 - 4 TIMES
22		:12	6/71·5A	LAYIN' ROCK UPON YOUR BRAIN F:09	40 - 5 TIMES
23	>	:04	6/71·6A	WNAP F	40 - 6 - TIMES
24		:07	6/71·5C	WNAP IS A FOUR LETTER WORLD	40 - 4 TIMES
25	—	:20	6/71·1A	--- IN INDIANAPOLIS, YEAH S-F :18	100 - 3 TIMES
26		:07	6/71·3	WNAP STEREO 93 F :05	40 - 4 TIMES
27		:12	6/71·2B	ELELECTRONIC MAMA, STEREO 93 :07 F ME	70 - 5 TIMES
28		:27	6/71·4	UNFILTERED MACHINE F-S :25	100 - 3 TIMES
29		:06	6/71·5B	STEREO 93 IS SAFER YEAH	40 - 5 TIMES
30		:02	6/71·6B	WANNAP! (WORLD'S SHORTEST STATION-BREAK)	40 - 10 TI
31	—	:07	6/71·1B	STEREO 93 F :05½	40 - 4 TIMES
32	:02	:28	6/71·6B	WANNAP talk over :06½/:26	100 - 3 TIMES
33		:16	6/71·1A	IN INDIANAPOLIS - LOGO S-S :15	70 - 4 TIMES
34		:13	6/71·2A	I AM SOUND · WNAP S :10	70 - 4 TIMES
35	:02	:21	6/71·5B	SULFURIC ACID ROCK :18 F	100
36		:16	6/71·6A	HERE'S 93 FOR ONE AND ALL	
37	:02	:33	6/71·4	THE BRAND NEW OLD TIME FLYIN' M	
38		:07	6/71·5C	REMEM :05	
39			6/71·1A	AL	
40					
41					

	Monophonic 2 Tracks			Stereophonic 4 Tracks*	
7½ ips	15 ips	1⅞ ips	3¾ ips	7½ ips	15 ips
1 hr.	30 min.	8 hrs.	4 hrs.	2 hrs.	1 hr.
1½ hrs.	45 min.	12 hrs.	6 hrs.	3 hrs.	1½ hrs.
2 hrs.	1 hr.	16 hrs.	8 hrs.	4 hrs.	2 hrs.
2 hrs.	1 hr.	16 hrs.	8 hrs.	4 hrs.	2 hrs.

*Double these times for 4-track monophonic recording.

THIS TAPE IS RECORDED: AT
☐ 1⅞ ☐ 7½
☐ 3¾ ☐ 15
☐ Monophonic (TRACKS)
☐ Stereophonic (TRACKS)

Few things influenced the sound of WNAP in the 1970s more than its purchase of two sets of jingle packages from producer Hugh Heller and creative man Dick Hamilton in Los Angeles.

Hugh Heller was a guy from Hollywood who was contacted by Jim Hilliard, who knew him personally. And he said, "Hugh, we have this little FM station here, and we want to do something really creative." And he said, "You mean, I can have some fun?" We used those for, good Lord, seven or eight years on the air. And to this day, I don't think I've played all the jingles. There were so many of them. I liked 'Klubeeda.' 'Klubeeda' was a six-second jingle. It just went 'Klubeeda! Dah, dah, dah! WNAP!' What the hell does that mean? It worked perfectly with our station. 'Cause you could play the dumbest record before it, some Gilbert O'Sullivan hit or something from Seventy-three into some cool album track, and nobody cared anymore. You made that song cool all of a sudden with this really weird jingle. —Buster Bodine

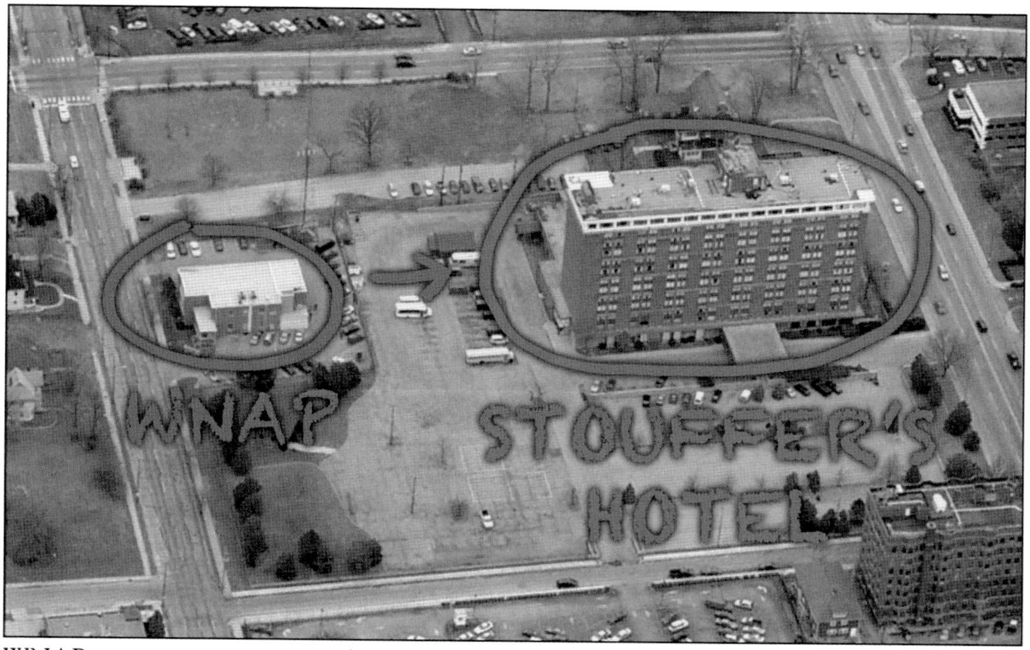

WNAP was generating national interest, and in 1973 George Johns was brought in to be the national program director for Fairbanks Broadcasting.

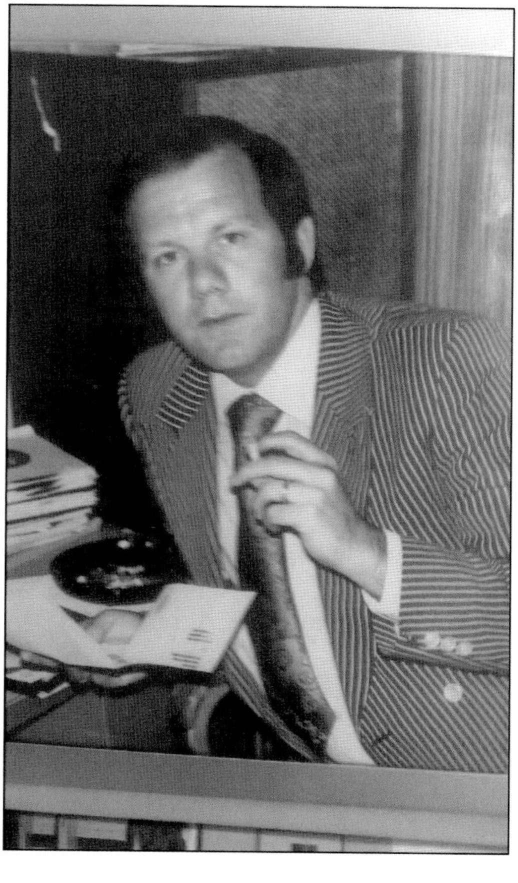

WNAP was probably one of the most listened to radio stations for about five years. Stouffer's Hotel, which was right across the alleyway from the "Two-Story Brick" at 28th and Illinois, would get constant barrages of program directors coming through listening to what we were doing because nobody understood us. And they had to sit there and listen. The fact is, you couldn't really figure out WNAP in one day. You had to listen for at least a week, maybe two weeks. And then it would start to sink in: "These guys, okay, alright, now we've got it." —Buster Bodine

It's a war of money. The war is the money. Who gets the money? Our war was to get more people listening to our radio station than listening to theirs. My job was to first fix WNAP. It had already evolved into a very esoteric, mostly album-oriented radio station. My job was to bring it back to the longer versions of hits. There are no ratings waiting in deep cuts. —George Johns

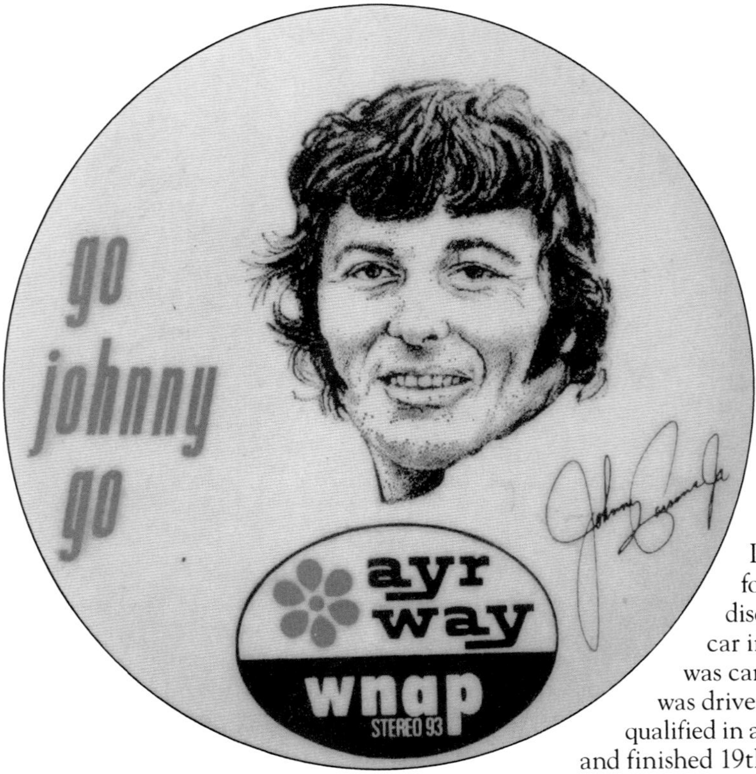

In 1975, WNAP joined forces with Ayr-Way, a local discount retailer, to sponsor a car in the Indianapolis 500. It was car no. 93—of course—and was driven by Johnny Parsons Jr. It qualified in a respectable 12th position and finished 19th.

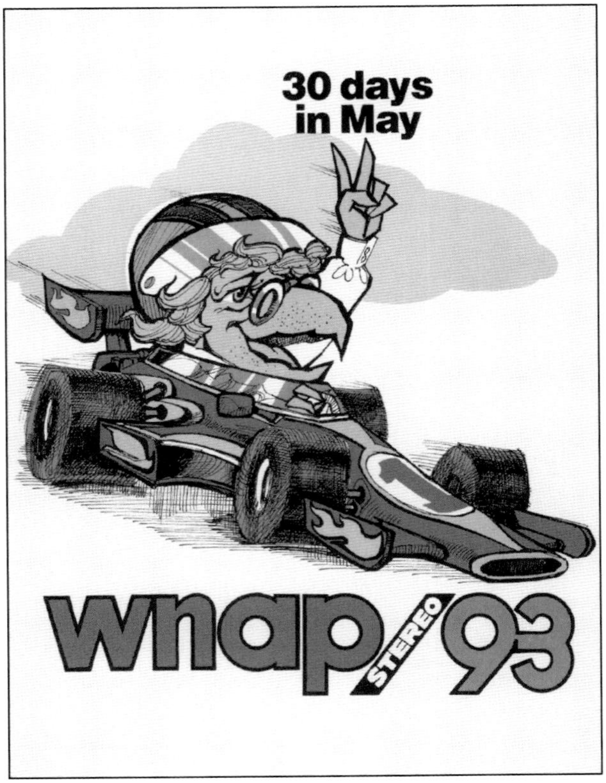

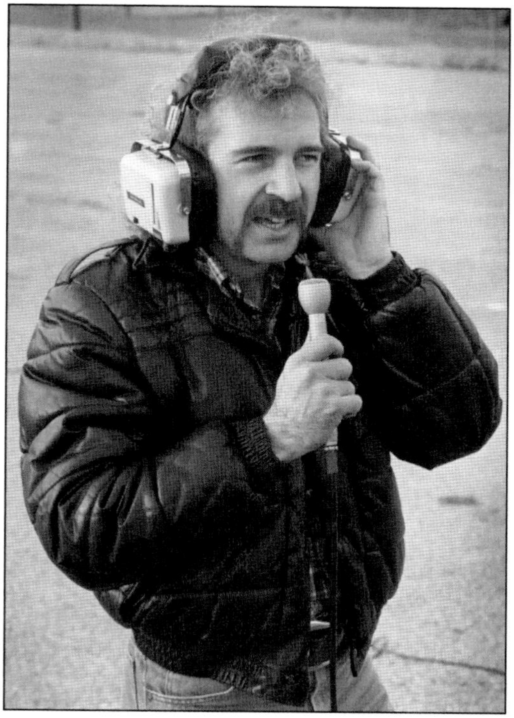

Steve "Fast Freddie Fever" Reynolds was an at-large air personality on WNAP but would often drop in for Cris Conner's morning show. Freddie is best remembered for his strangely dead-on impressions of Tom Snyder, Mr. Rogers, and Julia Child. He was a true ambassador for the station, being seen frequently at various nightspots around Indianapolis.

Symbol of WNAP FM is "The Buzzard".

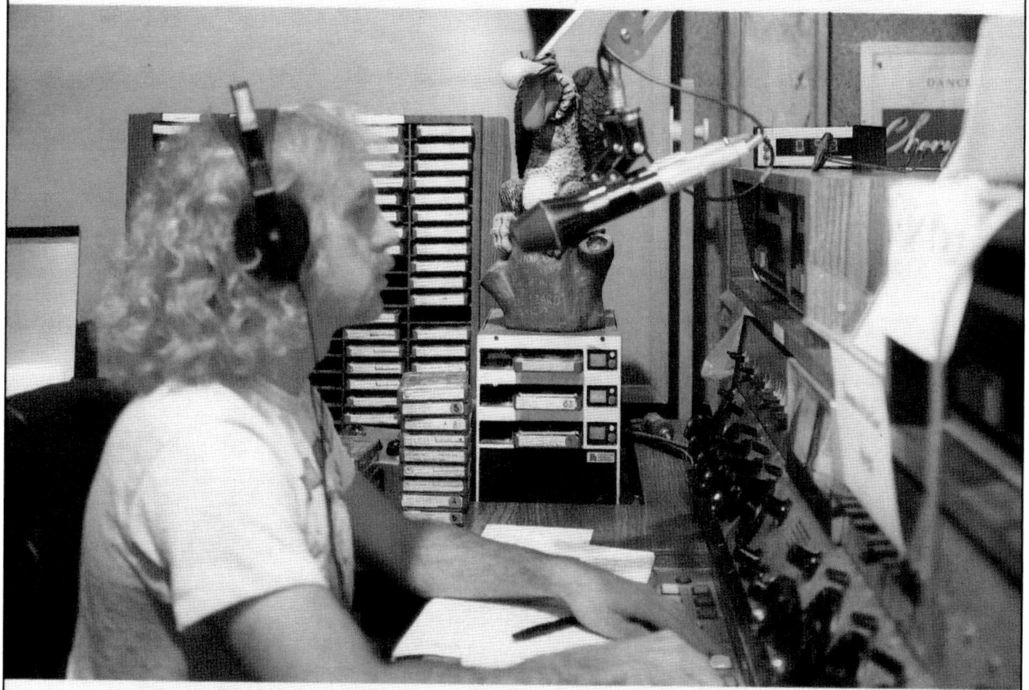

Air personality, Freddie Fever, and WNAP mascot, "The Buzzard", on the air for Radio Indiana's No. 1 "Rocker".

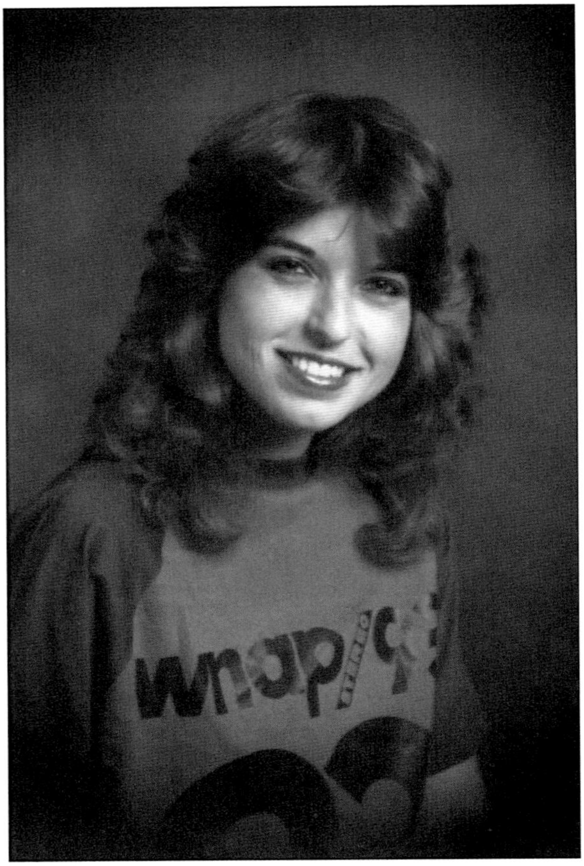

Ann Craig-Cinnamon started out in the newsroom at WIFE and moved to WNAP in 1979 to fill out Cris Conner's morning show when Tom Cochrun left to do a television show called *PM Magazine*. Ann was a trailblazer, as it was unusual for a woman to be such a prominent part of a morning show at that time. Her success at WNAP helped launch a 30-year career in Indianapolis broadcasting, most of which would be hosting morning shows.

This sharply dressed crew includes, from left to right, Fast Freddie Fever, Cris Conner, and Ron Below holding Ann Craig-Cinnamon.

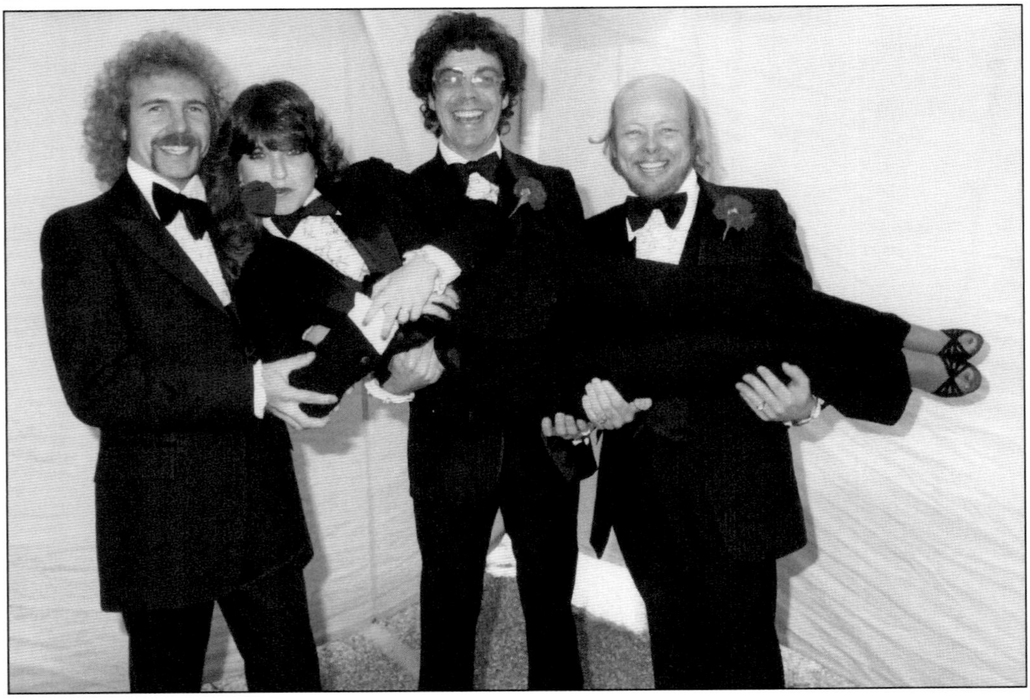

Four

THE UNDERGROUND AND RADIO FREE NAPTOWN

Modern bootleg radio in Indianapolis began in 1963 when Arlington High School students Barry Jansen and Don Worsham set up a small, unlicensed radio station near 46th Street and Arlington Avenue with the call letters WMRT. The next year, Steve Ross and Alan Taylor launched WISK on the west side of the city. Then, Broad Ripple High School students Steve Everitt and Neil Bolding fired up another bootleg station, WSSE. The three stations eventually merged, combining their equipment and ideas to create WMRT in March 1966.

They were inspired by what they perceived as stale programming coming from WIFE, the most popular station in town. There was no hint on Indianapolis radio of the new progressive rock coming out of West Coast cities. Sure, there would be numerous threats from other broadcasters and narrow escapes from FCC storm troopers, but so what? That was part of the adrenaline rush of having a bootleg radio station. These guys were having fun.

In 1968, Steve Everitt hatched a plan to develop Radio Free Naptown, or WRFN, and broadcast it at 1110 kilohertz—the "Heavy Eleven"— with an alternative underground music format. By September 1970, the station was flirting with major success, so WFBM program director Jack Sorbi became interested in having them program WFBM-AM and -FM over the weekends. The deal was made with Sorbi and the bootleggers, but WFBM station manager Jerry Chapman shot it down.

The WRFN transmitter bounced around Indy's north side a couple of more times, but in March 1972 the jig was up. The FCC, FBI, Department of Justice, and Indianapolis Police Department shut things down and hauled off the culprits, who pled no contest to charges and were fined. It was the end of seven years on the air by one of the nation's most notorious bootleg rock radio stations.

WIFE was a terrible radio station because all you heard were these smiling voices. You didn't hear a lot of musical variety. There was a very, very limited playlist. And contests are great, but as a listener they become an annoyance. —Don Worsham

At WIFE, all the guys were pukers. A "puker" is a guy who talks like this all the time because he's about [coughs] ready to throw up. That's pretty much a puker. —Tom Woody

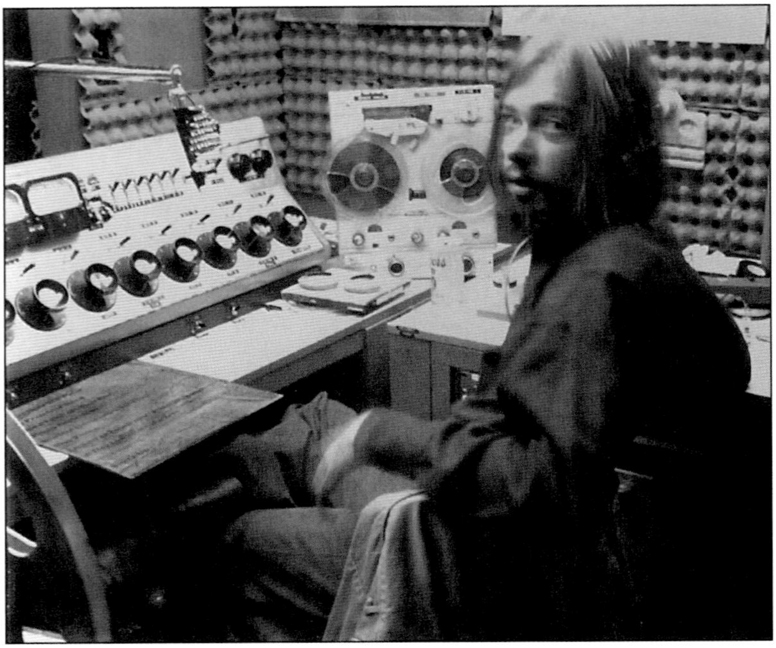

Tom Woody sits at the audio board of bootleg station Radio Free Naptown, WRFN, which was in no way related to WNAP. WNAP had been a disappointment to the bootleg radio crowd for not living up to its potential.

Radio Free Naptown was a bootleg radio station operating at odd times that was providing an alternative both musically and culturally that didn't exist at that point in time in Indianapolis. A bootleg radio station is one that is unlicensed because licensing requirements were very strict at that point. —Neil Bolding

Our bootleg stations were a reaction to a lack of programming on current radio at that time in Indianapolis. —Neil Bolding

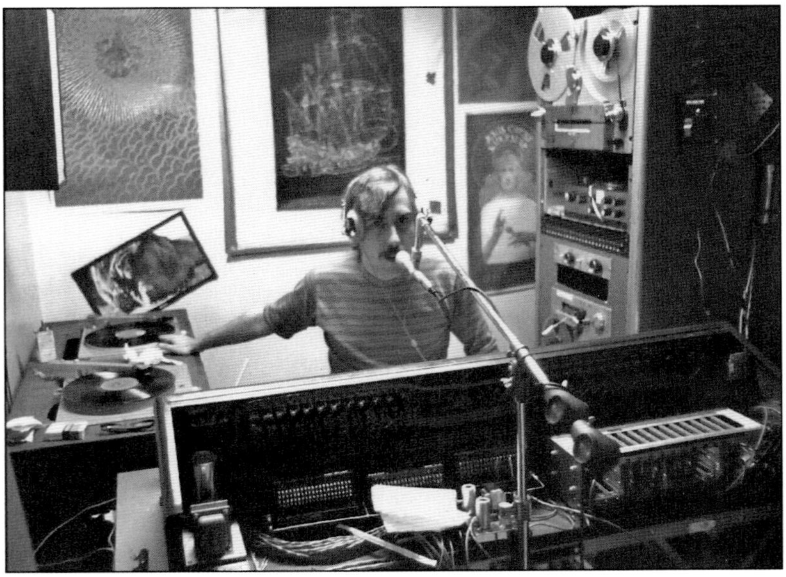

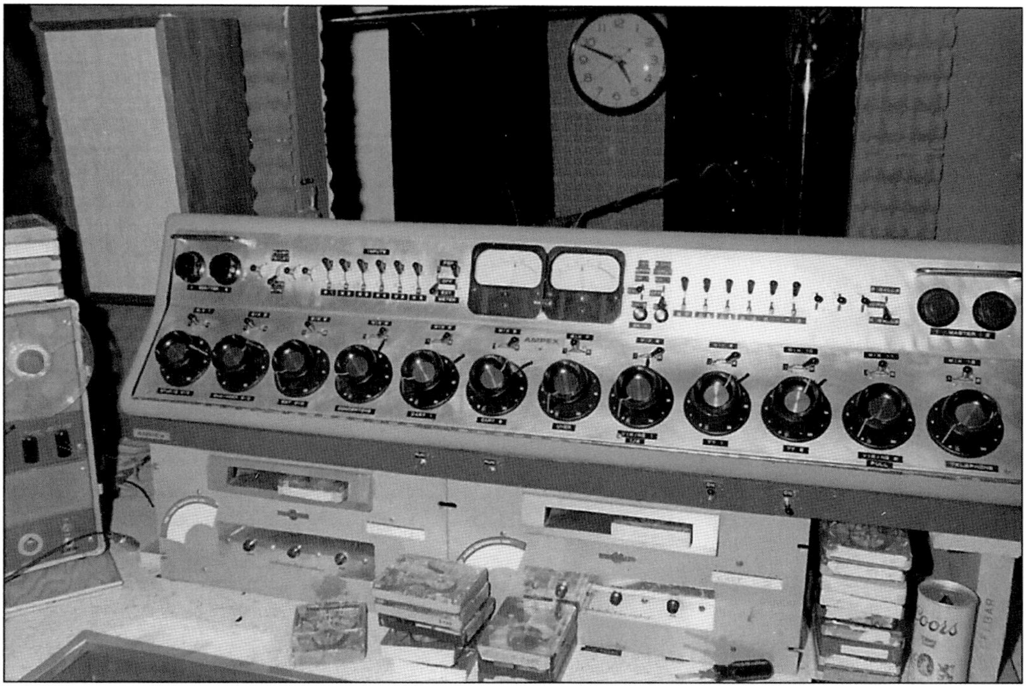

Most of the WRFN staff was made up of broadcasting professionals.

This was not, you know, a record player and a handheld microphone. This was broadcast-quality equipment because that was what they were used to working with. —Neil Bolding

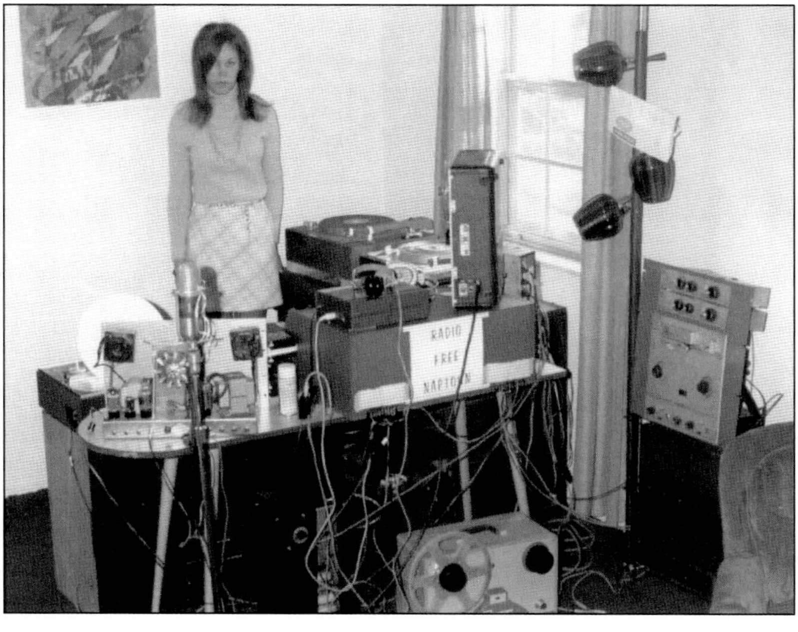

As we started to play these cuts, we opened up what we called a phone request line. We'd send somebody to a phone booth—there were phone booths at that point in time—and give out the number, and people would call. And that person would write down requests and drive back to the station, and we'd have a request list. —Neil Bolding

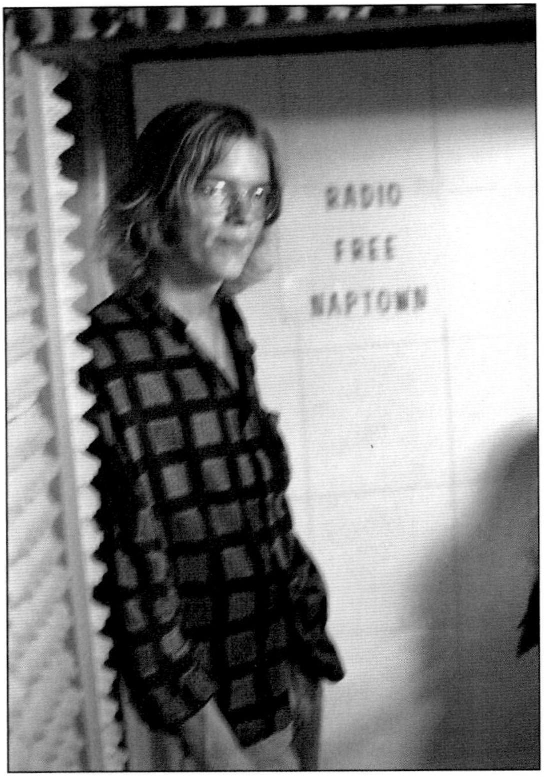

Bizarre on-air monikers were common with the bootleg radio operators. Tom Woody was known as "Thunderthroat." Doug Barry (pictured) took on the name "Sunshine." The WRFN-AM transmitter and tower were moved to the backyard of a house on College Avenue (below).

Radio Free Naptown was the most notorious illegal bootleg radio station in the country. —Tom Woody

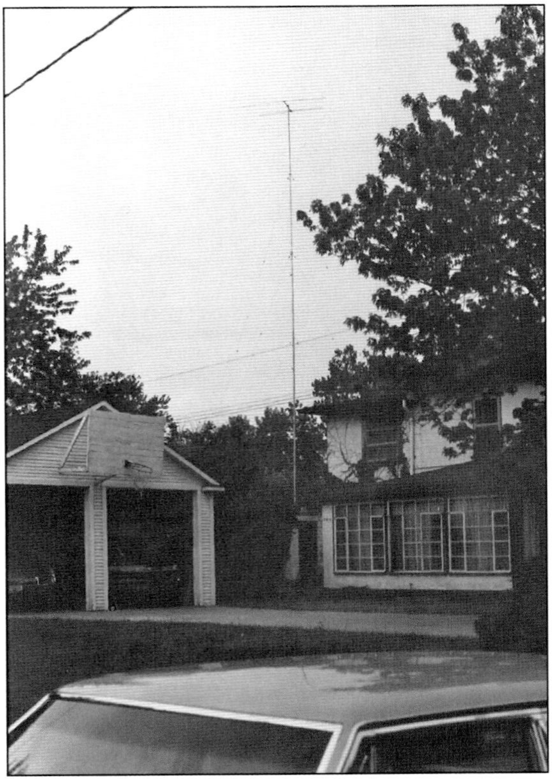

There were a number of rumors about how Radio Free Naptown managed to avoid detection. Some of the better ones were that we were in a van. Another one was the antenna was hidden in a hose that was laying in the front yard someplace. I can say with a certain degree of confidence that if anyone had seriously wanted to find Radio Free Naptown when we were on the air with AM, you could have found us. —Don Worsham

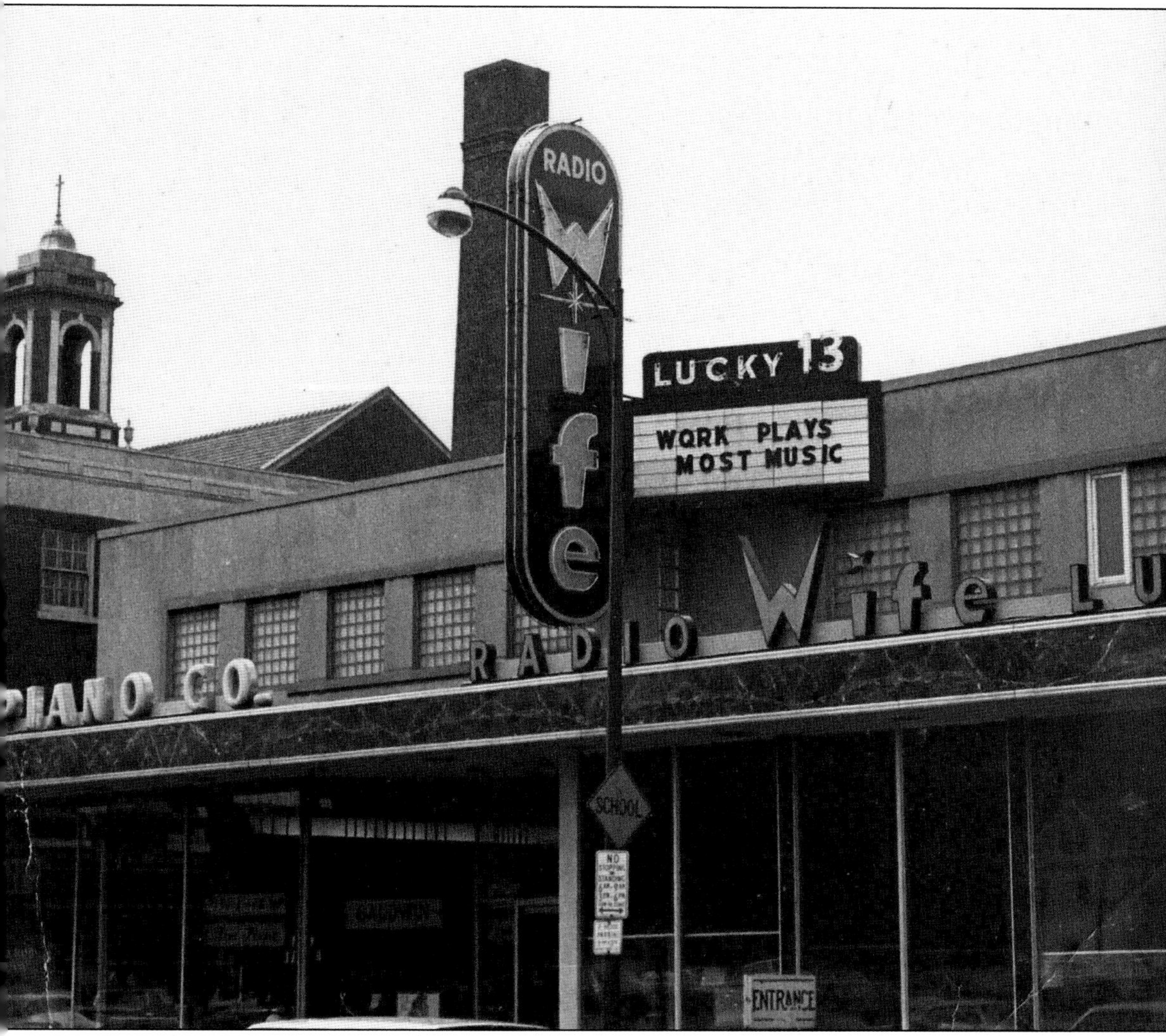

Here is a direct assault on WIFE by pranksters at WQRK.

We "punked" WIFE as a bootleg radio station in the winter of Sixty-seven. Our station was called WQRK at the time. We got up there at midnight to the WIFE marquee, took down all their signage, and put up "WQRK plays most music." And that punking of WIFE was pretty funny at the time at midnight. I went back the next afternoon at four, what's the marquee say? "WQRK plays most music." —Don Worsham

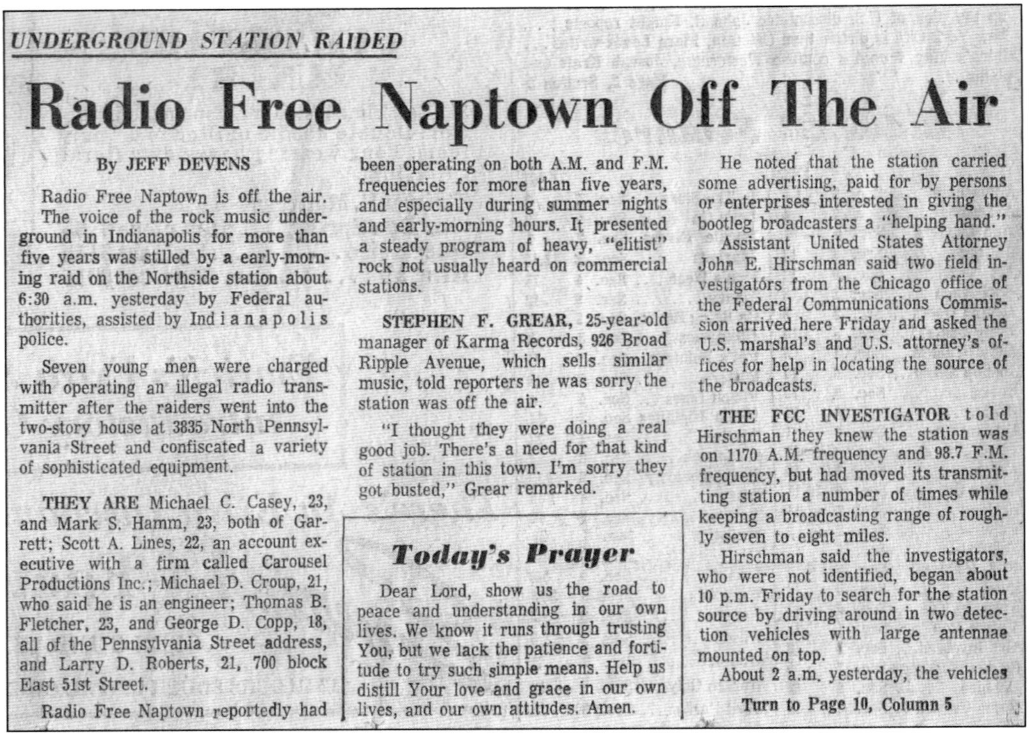

UNDERGROUND STATION RAIDED

Radio Free Naptown Off The Air

By JEFF DEVENS

Radio Free Naptown is off the air. The voice of the rock music underground in Indianapolis for more than five years was stilled by a early-morning raid on the Northside station about 6:30 a.m. yesterday by Federal authorities, assisted by Indianapolis police.

Seven young men were charged with operating an illegal radio transmitter after the raiders went into the two-story house at 3835 North Pennsylvania Street and confiscated a variety of sophisticated equipment.

THEY ARE Michael C. Casey, 23, and Mark S. Hamm, 23, both of Garrett; Scott A. Lines, 22, an account executive with a firm called Carousel Productions Inc.; Michael D. Croup, 21, who said he is an engineer; Thomas B. Fletcher, 23, and George D. Copp, 18, all of the Pennsylvania Street address, and Larry D. Roberts, 21, 700 block East 51st Street.

Radio Free Naptown reportedly had been operating on both A.M. and F.M. frequencies for more than five years, and especially during summer nights and early-morning hours. It presented a steady program of heavy, "elitist" rock not usually heard on commercial stations.

STEPHEN F. GREAR, 25-year-old manager of Karma Records, 926 Broad Ripple Avenue, which sells similar music, told reporters he was sorry the station was off the air.

"I thought they were doing a real good job. There's a need for that kind of station in this town. I'm sorry they got busted," Grear remarked.

Today's Prayer

Dear Lord, show us the road to peace and understanding in our own lives. We know it runs through trusting You, but we lack the patience and fortitude to try such simple means. Help us distill Your love and grace in our own lives, and our own attitudes. Amen.

He noted that the station carried some advertising, paid for by persons or enterprises interested in giving the bootleg broadcasters a "helping hand."

Assistant United States Attorney John E. Hirschman said two field investigators from the Chicago office of the Federal Communications Commission arrived here Friday and asked the U.S. marshal's and U.S. attorney's offices for help in locating the source of the broadcasts.

THE FCC INVESTIGATOR told Hirschman they knew the station was on 1170 A.M. frequency and 98.7 F.M. frequency, but had moved its transmitting station a number of times while keeping a broadcasting range of roughly seven to eight miles.

Hirschman said the investigators, who were not identified, began about 10 p.m. Friday to search for the station source by driving around in two detection vehicles with large antennae mounted on top.

About 2 a.m. yesterday, the vehicles

Turn to Page 10, Column 5

Finally, the jig was up.

Over time, there came the knowledge that there were people in Indianapolis that were looking to have the FCC come in and shut down the station. In March of 1972, the Federal Communications Commission, the FBI, and the Indianapolis Police Department raided the radio station. —Neil Bolding

When they got busted, they actually had the transmitter down near 38th and Pennsylvania, and it was a mailman who actually was the snitch. —Tom Woody

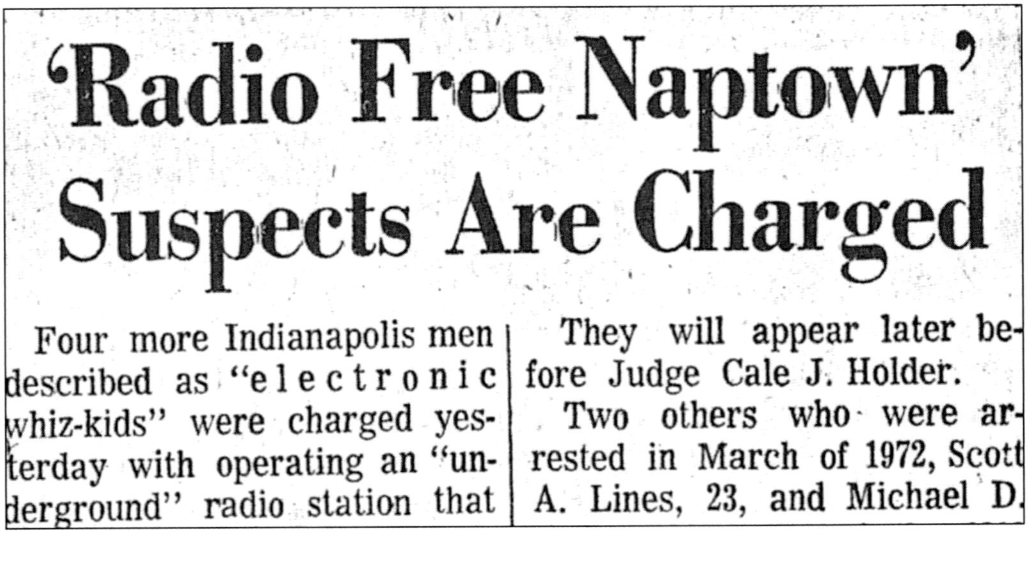

'Radio Free Naptown' Suspects Are Charged

Four more Indianapolis men described as "electronic whiz-kids" were charged yesterday with operating an "underground" radio station that

They will appear later before Judge Cale J. Holder.

Two others who were arrested in March of 1972, Scott A. Lines, 23, and Michael D.

Five

STIFF COMPETITION FROM WTLC, WNDE, AND WXLW

While the primary radio war in Indianapolis was between WIFE-AM/FM (Star Stations) and WIBC/WNAP (Fairbanks Broadcasting), there were other players entering the rock radio arena.

On January 22, 1968, WTLC signed on the air as the first full-time R&B station in Indianapolis. Further, they had taken the plunge and ventured into the FM band a full six months ahead of WNAP. Before WTLC, Indy's African American radio audience was divided. There was WGEE, which had gone along for years splitting its airtime between country and western during the day and R&B the rest of the time. Then there was WIFE, which provided a steady diet of Motown, Stax, Volt, and whatever else was finding its way onto the Top 40 charts. When WTLC signed on, it made a dent in both WGEE and WIFE's audiences.

In 1973, WFBM-AM was the oldest radio station in Indianapolis. It was then sold, and the new owners retooled it completely to compete with WIFE. The call letters changed to WNDE (pronounced "windy"), and heavy-hitter programmers such as Bill Hennes (WMAQ and CKLW) and Chris Bailey were brought in to ratchet up the energy. The energy was so intense, a deejay would end the shift drenched in sweat.

WXLW first signed on the air on August 18, 1948, and operated as a middle-of-the-road station until Bill Shirk took control in 1972 as station manager. In 1974, he bought the place. WXLW became Shirk's playground, and it reflected his style and personality. He played the top hits and presented outlandish contests to listeners. A tireless worker and creative entrepreneur, Shirk turned his little daytime-only AM station into a competitor for Indy's rock radio scene.

In the early '70s, WNAP and WIFE weren't broadcasting in a vacuum. —Al Stone

We had had a consistent diet of the good guys over at WIFE. As a matter of fact, they did play black music, but it was in Top 40 format as one could recognize, or one could remember. And it was really great. But this [WTLC] was just unbelievable, to get that consistent diet, and then get it 24/7. Oh, it was absolutely amazing. —Al "The Bishop" Hobbs

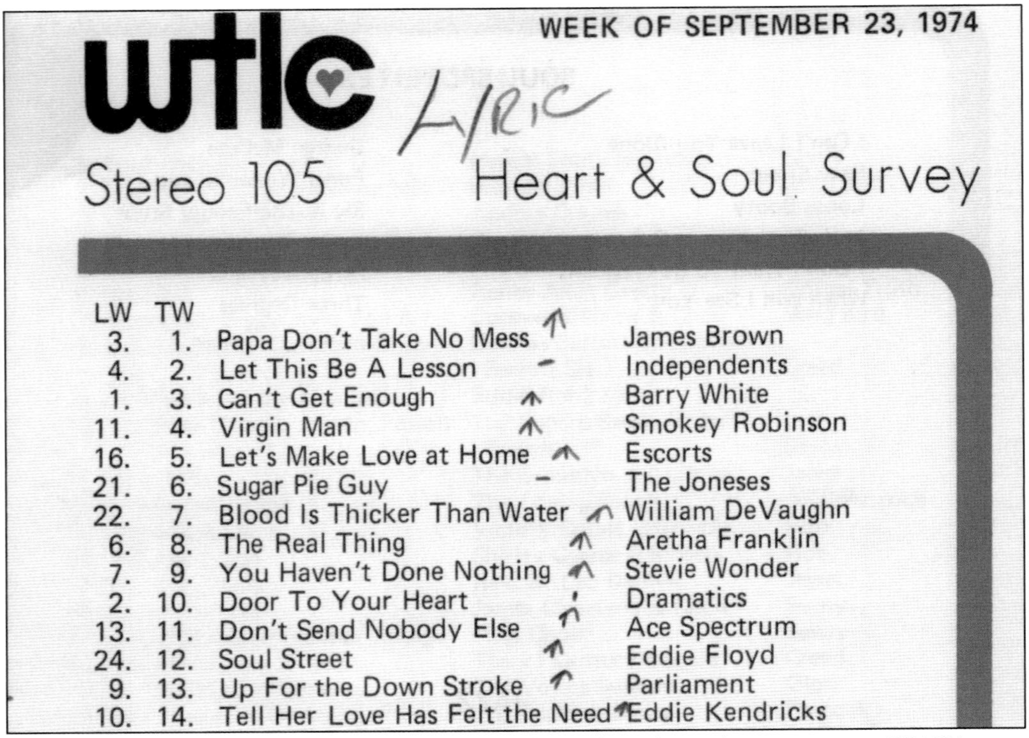
WEEK OF SEPTEMBER 23, 1974

wtlc

Stereo 105 Heart & Soul Survey

LW	TW	Title	Artist
3.	1.	Papa Don't Take No Mess	James Brown
4.	2.	Let This Be A Lesson	Independents
1.	3.	Can't Get Enough	Barry White
11.	4.	Virgin Man	Smokey Robinson
16.	5.	Let's Make Love at Home	Escorts
21.	6.	Sugar Pie Guy	The Joneses
22.	7.	Blood Is Thicker Than Water	William DeVaughn
6.	8.	The Real Thing	Aretha Franklin
7.	9.	You Haven't Done Nothing	Stevie Wonder
2.	10.	Door To Your Heart	Dramatics
13.	11.	Don't Send Nobody Else	Ace Spectrum
24.	12.	Soul Street	Eddie Floyd
9.	13.	Up For the Down Stroke	Parliament
10.	14.	Tell Her Love Has Felt the Need	Eddie Kendricks

The 105.7 FM frequency in Indianapolis had belonged to classical music station WAIV since 1961, prior to becoming WTLC in 1968.

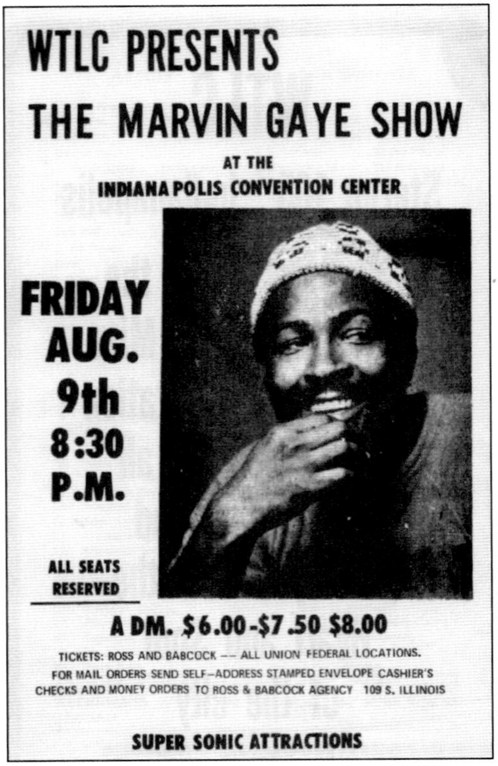

In January of 1968, a group of businessmen and political personalities in our community got together and put on the FM band a station called WTLC with tender loving care. I remember when I first heard the station, I just got so excited I couldn't believe it because it was a consistent 24-hour diet of great—as a matter of fact, the "Great 48"—black records in the country. —Al "The Bishop" Hobbs

WTLC's sign on in 1968 was the beginning of the erosion of events at WIFE. That continued into the Seventies on many other stations. —Russ Dodge

"Super Jay" Johnson was a top talent at WTLC and personified the growth of the station. Johnson came from WVON in Chicago, took over the afternoon-drive time, and would go on to become program director. His popular show featured visits by nationally respected African American celebrities on a regular basis.

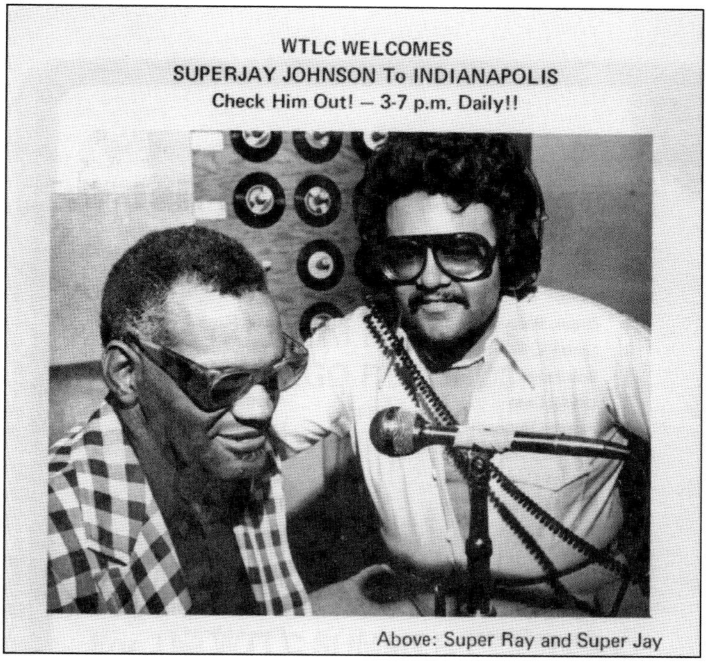

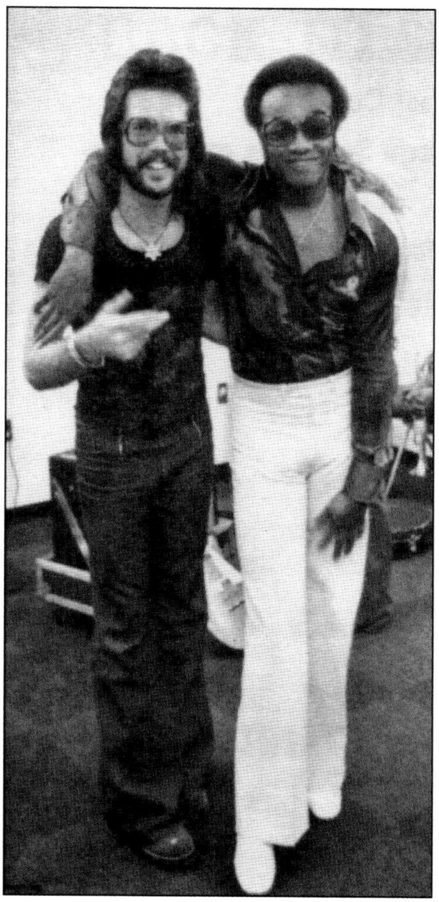

One of the original deejays at WTLC was Roger Holloway (left). Roger's career included being music director for WTLC, a member of the band Pure Funk, and a Christian evangelist.

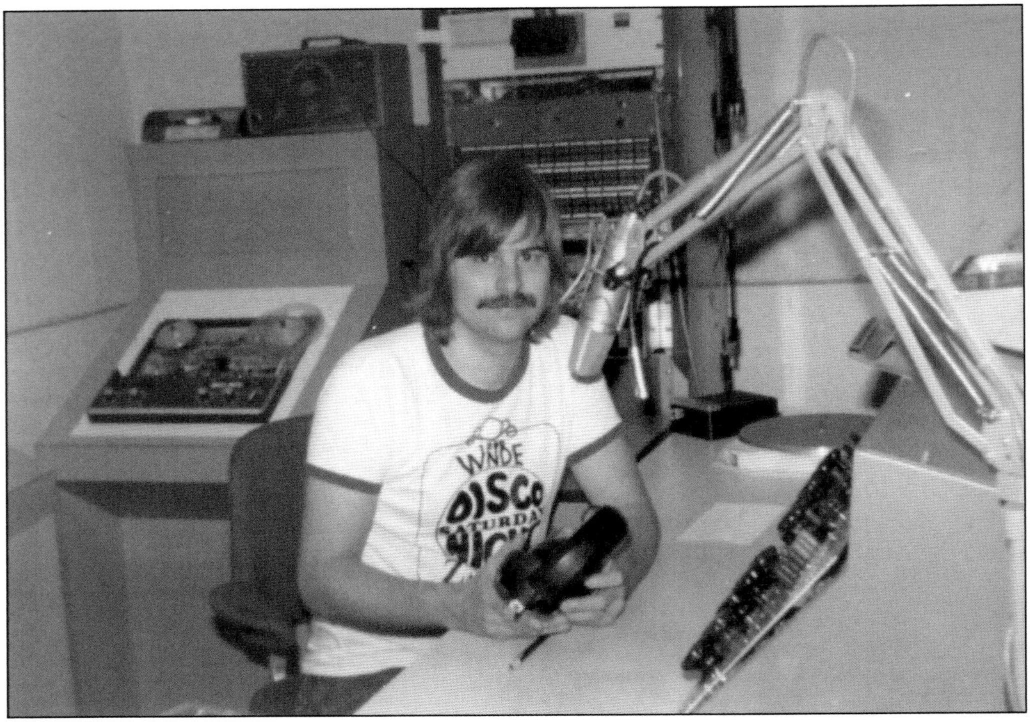

Mike O'Brien (above) and Scott Wheeler (below) are two of several deejays who moved to WNDE from WIFE.

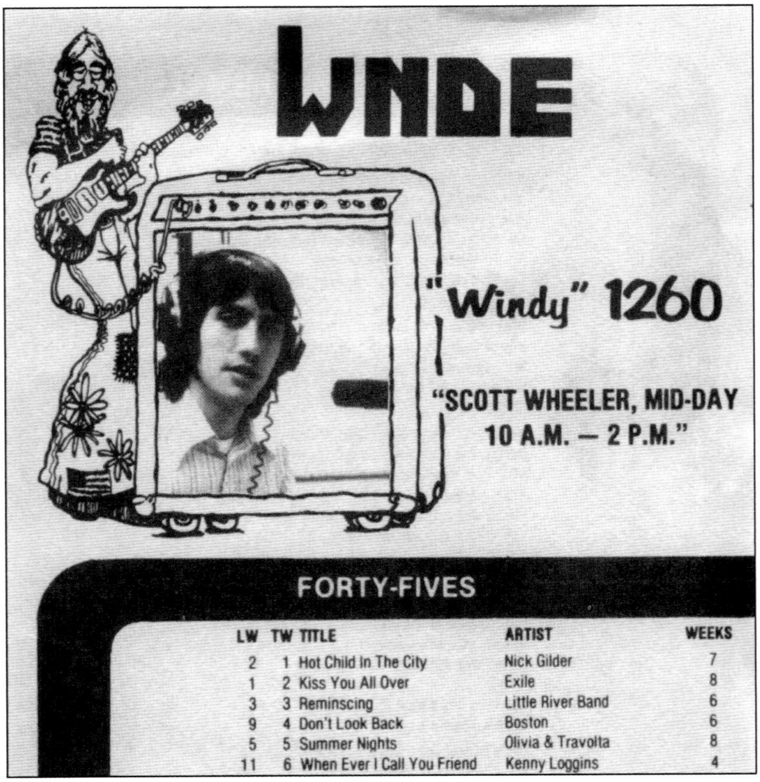

These two radio stations played pretty much the same thing, so it was all in the presentation, and the signal in some cases, because WIFE had a terrible signal at night, whereas WNDE was a little bit better.
—Steve Cooper

WNDE on 1260, and WIFE on 1310. So, the two radio stations are practically hugging each other on the dial!
—Art Vuolo Jr.

The call letters "NDE" were intended to represent "Indy."

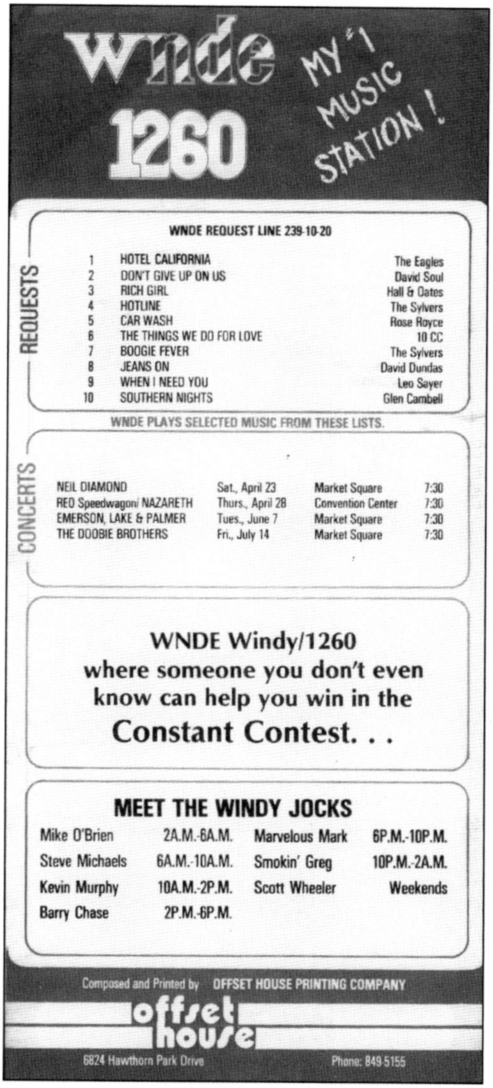

WNDE came to market with a group of very talented people, a very large collection of music, and a very, very good, tight format. —Steve Cooper

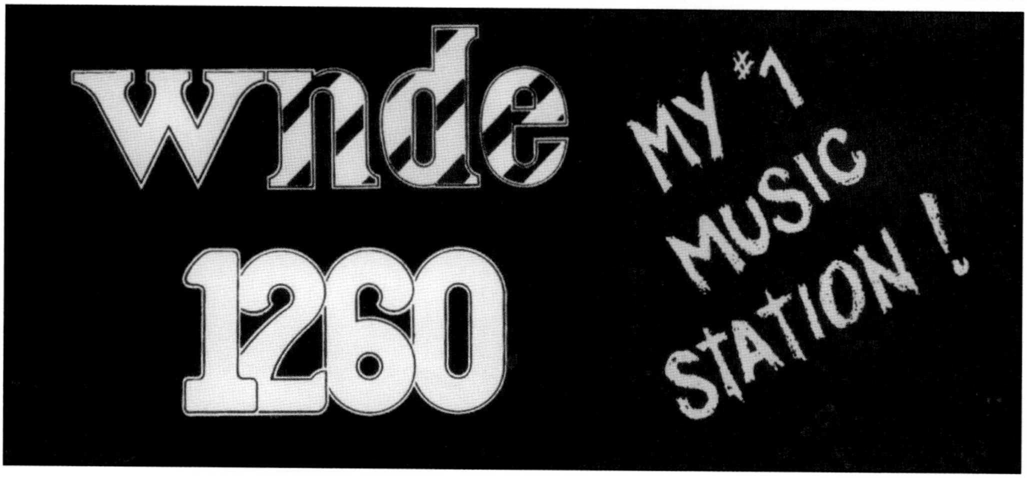

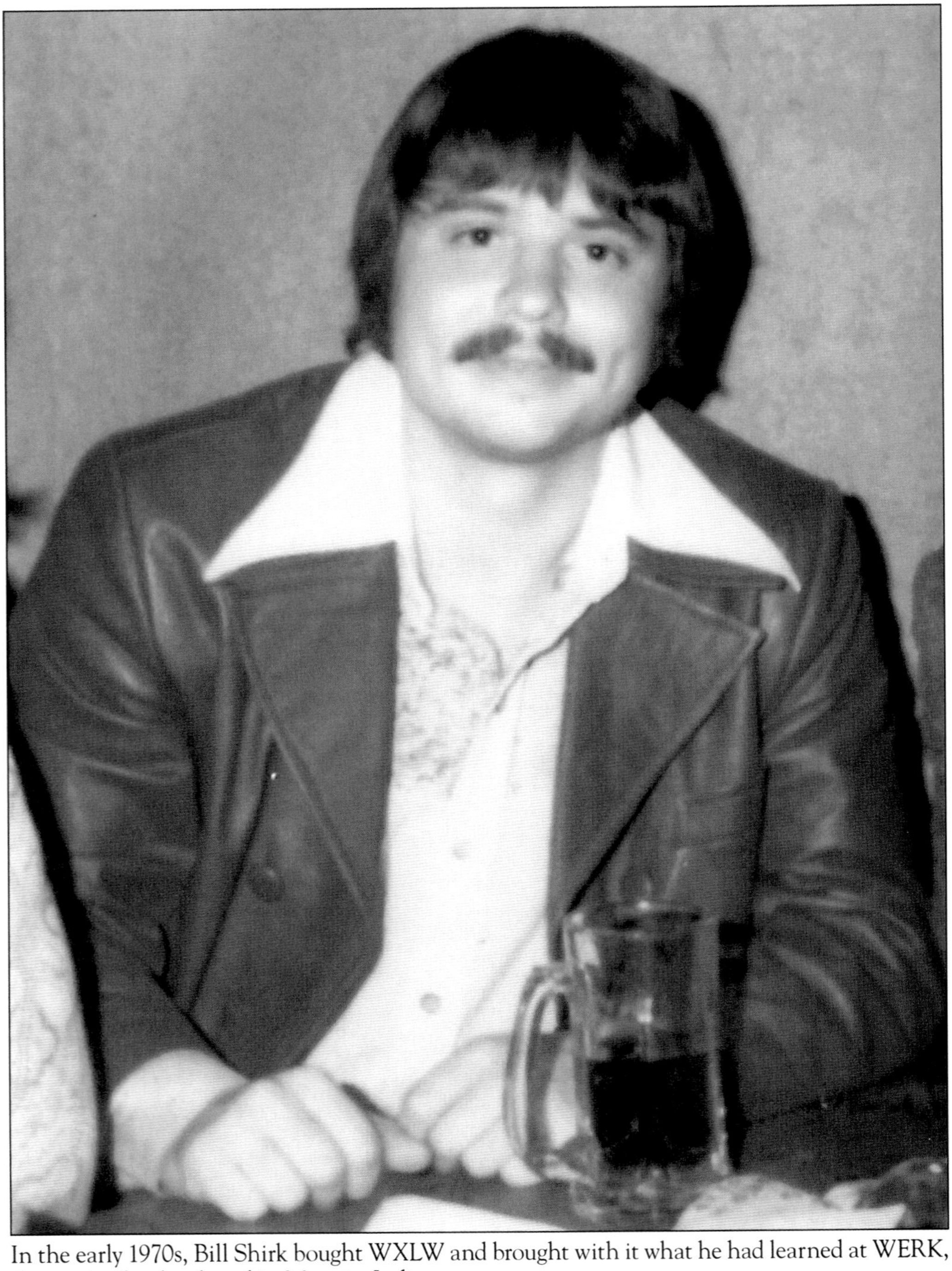

In the early 1970s, Bill Shirk bought WXLW and brought with it what he had learned at WERK, the station he developed in Muncie, Indiana.

Bill Shirk was very, very interesting when he took over WXLW. Many people wondered, "Is he going to make it?" Well, as gutsy as he is and was, he was going to make it. —Pat Kelly

I would say Bill Shirk, the best way I could describe him, is he's almost a carny type. He's all about the promotion, and the flash, and the whatever. —Ann Craig-Cinnamon

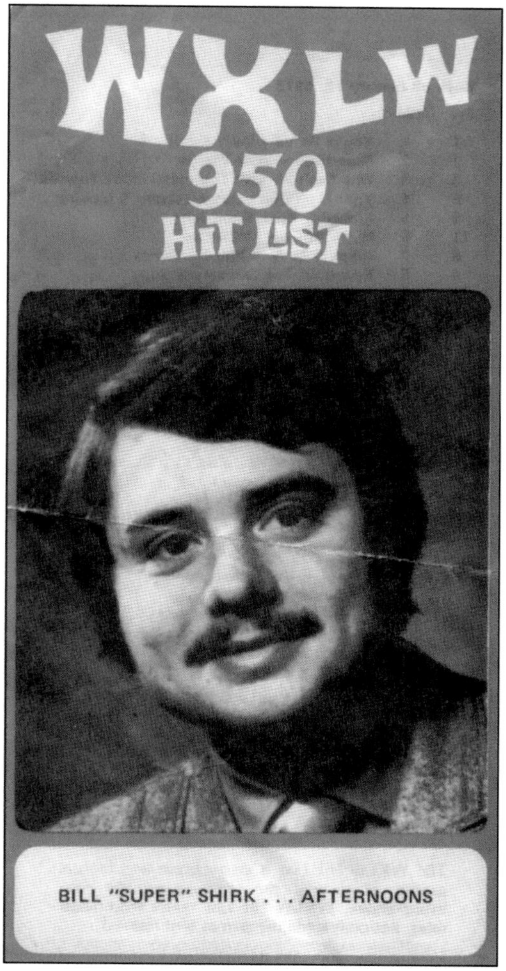

Music surveys were lists of the current hits compiled by a radio station, and WXLW created the best compared to other rock stations. Shirk pounded contests into the ears of the listeners constantly, and music surveys frequently featured them as well as the constant self-promotion of his show.

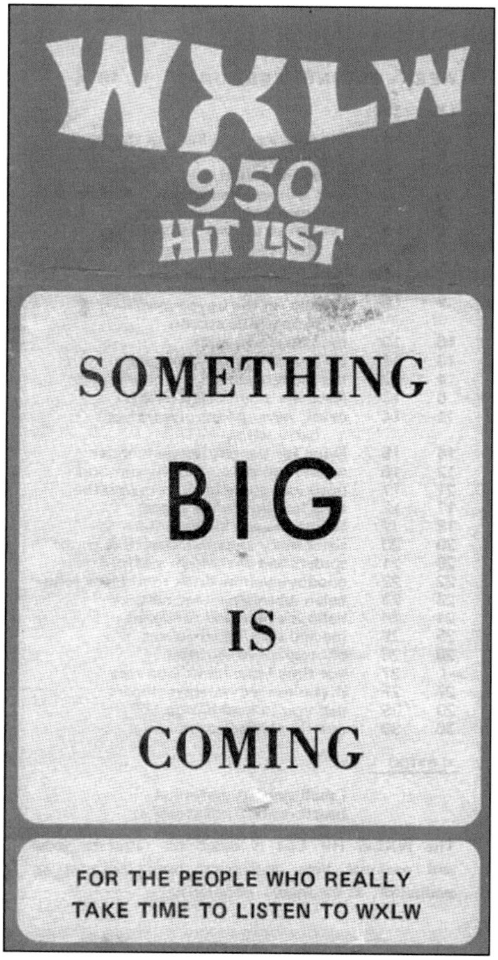

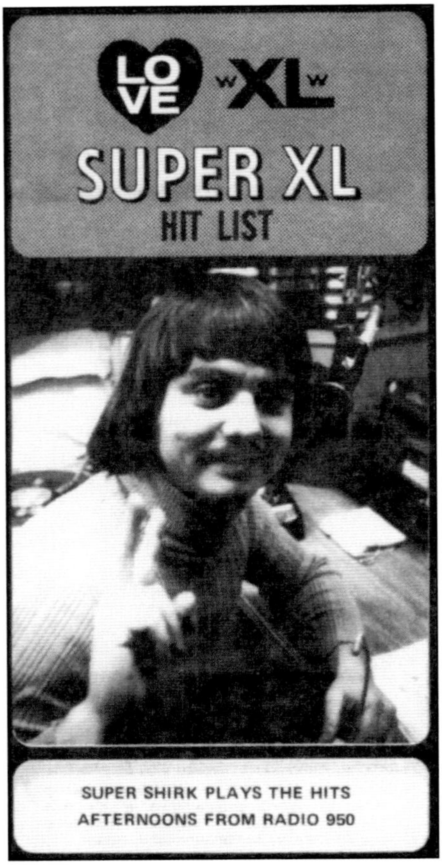

Bill Shirk's WXLW had restricted broadcast hours and a very low power transmitter.

WXLW's signal, they had a daytime station. They had to go off the air, you know, at sundown. And they couldn't come on until sunup. —Pat Kelly

For a daytime-only station, WXLW did very well. —Al Stone

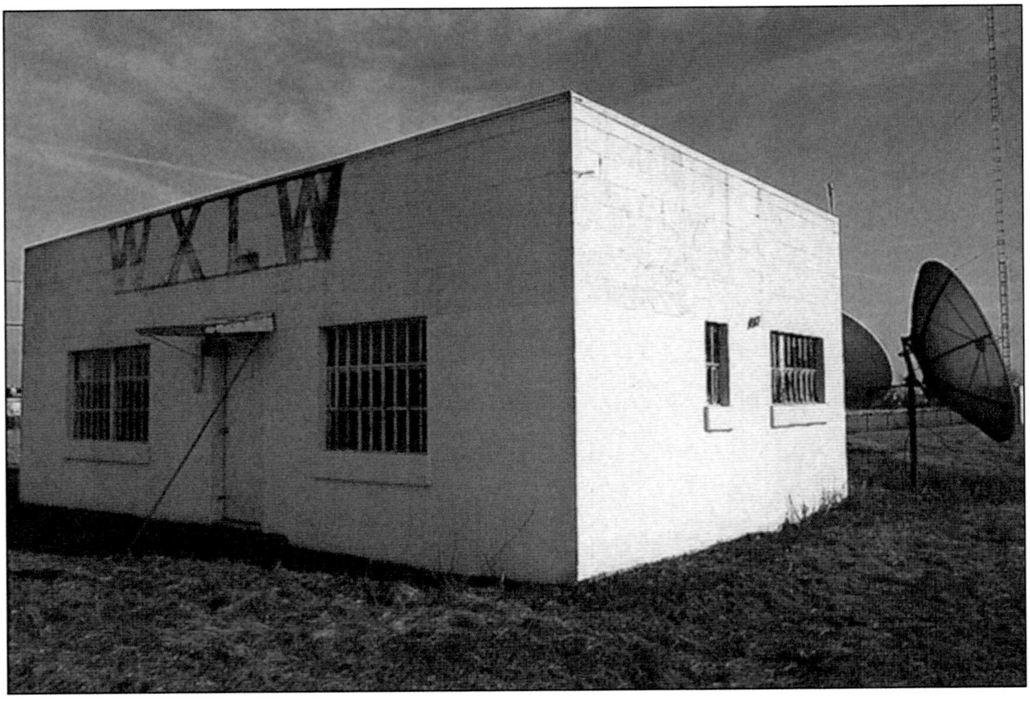

Six

The Decline of Lucky 13

Why did WIFE fall into the abyss? There are many reasons. What may be the biggest one was that AM radio was falling from grace. Sure, there would always be markets with an AM station leader, but it became a tough world out there for most AM broadcasters. Once the public wised up about FM radio and started to enjoy the clarity of the signal, with stereo sound and virtually no static, FM radios flew off the shelves and into pretty much every car on the highway.

WIFE was getting squeezed not only by competition from upstart FM stations, but from AM stations as well. WNDE not only gave them a run for the money, but by 1976 it was eclipsing them in the ratings. Whereas WIFE had no competition to speak of when it signed on in 1963, fifteen years later the station was getting beaten up in the ratings by stations "out-WIFEing" them.

Don Burden may have been a brilliant businessman and visionary, but he was difficult to work for. Sure, he may have slipped a tax-free C-note into an employee's palm for a job well done, but on a bad day it was said he might give a five-finger death punch to the kidneys if he was displeased with any staff—or so goes the legend. Few people stayed at WIFE for very long, and listeners like continuity in their deejays.

The fatal blow to WIFE came in the form of FCC accusations that Star Stations was slanting news coverage in favor of certain politicians who were running for office and that contests were being rigged. The combination of the FCC inquiry and Don's fiery temper and egotism was a deadly cocktail. It was speculated that at some point he said something to infuriate FCC commissioners, and they then chose to make an example of him.

Ultimately, the FCC took away all of Star Stations' licenses. Broadcasting had meant everything to Don Burden, and without his stations he died a broken man.

For the first time in WIFE's history in Indianapolis, they had a competitor. For the first time, there was a station that, over a period of time, began to outrate them. And they were no longer the number-one station. —Reb Porter

I think by the time they figured out something was happening it was too late. They were dead, they just didn't know it yet. It was a little harsh. Sure was fun though. —Jim Hilliard

WIFE

1440 N. MERIDIAN • INDIANAPOLIS, IND. 46202
TELEPHONE (317) 637-1375

KOIL OMAHA
KEFM OMAHA
KISN PORTLAND
WIFE INDIANAPOLIS
WIFE FM-STEREO INDIANAPOLIS

DON W. BURDEN, CHAIRMAN OF THE BOARD

STATEMENT TO ALL STAR STATION EMPLOYEES
(KOIL, KEFM-FM, KISN, WIFE, WIFE-FM)

The FCC today reversed Administrative Law Judge Naumowicz's decision in which he recommended that the FCC renew the licenses of all our Stations (with the exception of WIFE-AM, where a competing application had been filed).

As you know, the charges against us were brought by former disgruntled employees more than five years ago, of which we were exonerated on all counts by the Administrative Law Judge. After thousands of hours of testimony and hundreds of exhibits, the Administrative Law Judge exonerated us on all counts and was satisfied that all of our Stations had been operated in the public interest and recommended that our licenses be renewed. I have not, as yet, seen a copy of the Commission's decision, but I, and our attorneys, feel confident that it is fraught with errors of both fact and law.

In the same way that we vindicated ourselves, the reputation of the Stations, and the fact that all of the Stations have served the public inteeest to the satisfaction of Administrative Law Judge Naumowicz, I feel confident that we will ultimately prove these same truths before an impartial court. We have an absolute right to appeal this outrageous, biased decision. I have instructed our attorneys to immediately proceed to take such an appeal to the courts, which, as you know, will be an extremely long, drawn-out process.

I am confident that the court will reinstate Administrative Law Judge Naumowicz's decision to renew our licenses. In the meantime, it's business as usual.

DON W. BURDEN
STAR STATIONS, INC.

This internal document show's WIFE's battle with the FCC in the early stages and represents the beginning of the end.

Just as Don Burden had experienced meteoric success in the mid-Sixties, in the Seventies when he fell and when WIFE collapsed, Burden fell hard. —Al Stone

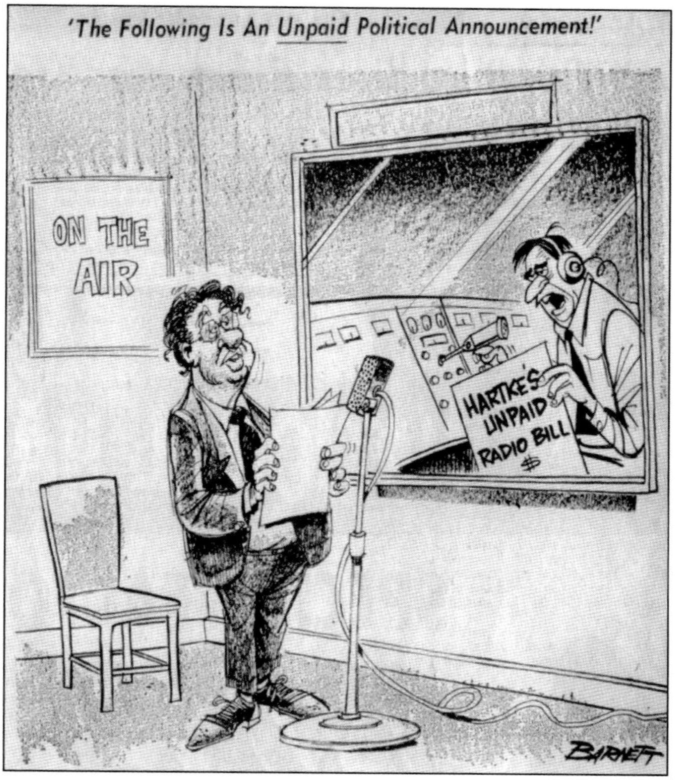

Among the charges by the FCC was that WIFE had provided Senator Vance Hartke with free and or discounted airtime and that the newscasts were being slanted in his favor for the 1964 election. Hartke was defeated in his 1976 reelection bid by Indianapolis mayor Richard Lugar in a landslide.

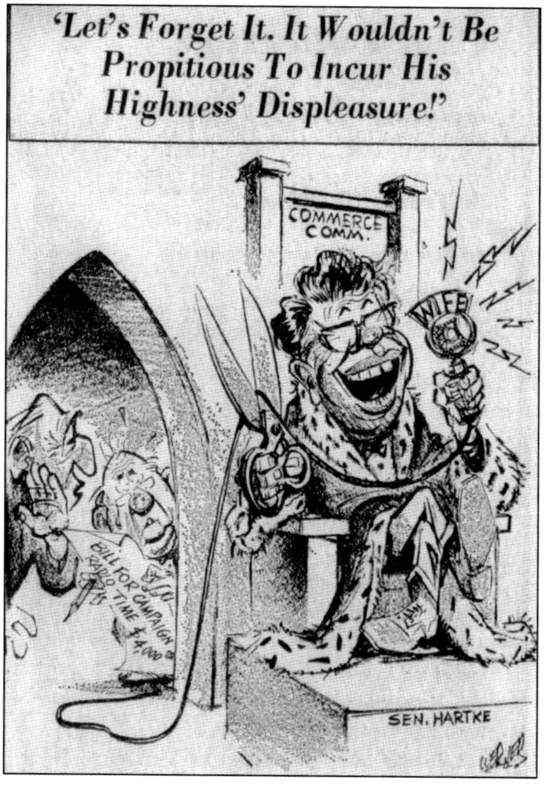

5202

BEFORE THE

Federal Communications Commission

WASHINGTON, D. C. 20554

In Re Applications of	)	
	)	
STAR STATIONS OF INDIANA, INC.	)	
	)	DOCKET NO. 16612
For renewal of licenses of Stations WIFE AM-FM, Indianapolis, Indiana	)	File Nos. BR-1144 BRH-1276
	)	
To: The Commission	)	

BRIEF OF STAR STATIONS OF INDIANA, INC.
IN SUPPORT OF
EXCEPTIONS TO INITIAL DECISION

Star Stations of Indiana, Inc., applicant in the above-captioned proceeding, by its attorneys, respectfully submits its Brief in Support of the Exceptions to the Initial Decision which are filed simultaneously herewith. The Initial Decision was released December 19, 1967 (FCC 67D-73); it proposed to deny the applications of Star Stations of Indiana, Inc. for the renewal of license of Stations WIFE and WIFE-FM, Indianapolis, Indiana.

INTRODUCTION

The Examiner's decision to deny the applications for renewal of license was based upon findings of station practices which, in the Examiner's language, were found to be "so offensive to the public interest" that there was "no reasonable alternative to the actions [so] recommended." The conduct that the Examiner referred to in making this judgment related to the issues in the case: (1) to the Mystery Santa Claus Contest

Did the FCC make an example of Don Burden?

I believe because he was very arrogant, he was very cocky, he was very "it's about me and I'm right, full speed ahead," I personally believe he alienated somebody in Washington. Whether it was the FCC itself, whether it was the commissioner of the FCC, whomever it was, and I think from that they made sure that he would never run a radio station or own a radio station thereafter. —Jack Marsella

THE INDIANAPOLIS STAR

WIFE-FM Ordered Off Air By FCC

Washington (AP)

FOUR RADIO stations were ordered to go off the air at 12:01 a.m. today by the Federal Communications Commission in what FCC officials said was one of the harshest actions the commission has ever taken.

The stations, owned by Star Broadcasting Co., lost last-minute appeals to slanting news broadcasts, and giving some political candidates air time without charging them.

On one occasion, the FCC said, Burden planned to pay off Vancouver officials for a zoning change for WISN. But the change was made "fortuitously" before any money changed hands, the FCC said.

AN FCC spokesman said the agency

L160

(FCC)

URGENT

(WASHINGTON)---STAR BROADCASTING HAS LOST A LAST-DITCH EFFORT TO AVOID SHUTTING DOWN RADIO STATIONS IN VANCOUVER, WASHINGTON, OMAHA AND INDIANAPOLIS AT 12:01 A-M TOMORROW.

TWO JUDGES ON THE U-S CIRCUIT COURT OF APPEALS IN WASHINGTON DENIED A REQUEST THAT AN ORDER HANDED DOWN EARLIER IN THE DAY BY THE FEDERAL COMMUNICATIONS COMMISSION BE BLOCKED.

THE STATIONS -- ALL OWNED BY BROADCASTER DON BURDEN -- ARE K-I-S-N IN THE PORTLAND, OREGON, AREA...W-I-F-E-FM AT INDIANAPOLIS, AND K-O-I-L AND K-E-F-M-FM AT OMAHA, NEBRASKA.

THE FEDERAL COMMUNICATIONS COMMISSION HAS ORDERED THE STATIONS OFF THE AIR AT 12:05 A-M LOCAL TIME TOMORROW. BURDEN HAS ASKED FOR MORE TIME IN WHICH TO TRY TO OFFSET AN F-C-C ORDER DECLINING TO RENEW HIS BROADCAST LICENSES.

BURDEN SAYS ANOTHER REASON TO AVOID DARKENING THE STATIONS IS TO PRESERVE THE JOBS OF ABOUT 100 EMPLOYES.

THE COURT OF APPEALS HAD AFFIRMED THE DENIAL OF A LICENSE LAST DECEMBER. THE LAWYERS TOLD THE COURT TODAY THAT THE STATIONS SHOULD NOT GO OFF THE AIR UNTIL THE F-C-C APPROVES APPLICATIONS FOR INTERIM LICENSES TO TAKE OVER THE STATIONS UNTIL A PERMANENT LICENSEE CAN BE FOUND.

THE ATTORNEYS SAID CLOSING THE STATIONS WOULD DENY THE THREE CITIES VARIETY IN BROADCASTING UNTIL INTERIM OPERATORS ARE LICENSED.

THE F-C-C HAS FILED PAPERS OPPOSING THE REQUEST FOR AN EMERGENCY INJUNCTION AGAINST ENFORCING ITS ORDER. THE COMMISSION SAYS IT WILL ACT QUICKLY ON INTERIM OPERATING APPLICATIONS, SUGGESTING THERE WOULD BE NO UNDUE DELAY IN REOPENING THE STAR BROADCASTING FREQUENCIES.

THE COMMISSION SAYS STAR ALREADY HAS BEEN GRANTED 90 DAYS TO PREPARE FOR CLOSING THE STATIONS SINCE THE SUPREME COURT TURNED DOWN BURDEN'S APPEALS OF THE LICENSING DENIALS.

UPI 09-01 03:40 PCD

Don Burden's lawyers kept the matter tied up in appeals for years. On May 31, 1976, Don Burden lost his licenses, and he was out of the radio business.

the law on this thing, and that's all there is to it."

Supreme Court Downs Star

The **Supreme Court**, Monday, left standing an FCC decision denying license renewals to several radio stations owned by **Star Broadcasting.** The **U.S. Appeals Court**, in Washington, affirmed the Februrary, 1975 FCC action, and the high court refused to hear an appeal which has now ended the case. The **FCC** had denied license renewals for **WIFE AM** and **FM** Indianapolis, **KOIL AM** and **FM** Omaha and **KISN AM** in Vancouver, Washington.

The **FCC** had found that **Don W. Burden**, who owns more than **92%** of **Star**, participated in actions that led to **WIFE** giving free air time to **Senator Vance Hartke** in the **1964** election campaign. The Commission also found Burden tried to use KISN to give one sided news coverage in favor of **Senator Mark Hatfield** during his **1966** campaign.

WIFE Fight Dropped By Burden

By RICHARD K. SHULL

Broadcaster Don Burden has abandoned his 12-year fight with the government to maintain control of radio station WIFE-AM here and will surrender the $3 million property to a group of local businessmen.

The sudden turn in the long-playing case came yesterday when Burden filed a motion with the U.S. Court of Appeals in Washington to cease his legal battle to retain control of the station.

Burden's move came after protracted bargaining with attorneys for the group that for six years has been seeking to take Burden's license for WIFE.

Burden's fight went up to the doors of the U.S. Supreme Court, which refused to hear the case.

This odd story appeared in the *Indianapolis News* days before the end. It serves as a curious footnote to Don Burden's broadcasting legacy.

Tee-Zee, Cee-Bee

Predictably, WXTZ here has screamed foul to the FCC over a request by Don Burden to change the call letters of WIFE-FM to WXCB.

"Phonetically and rhythmically" they're too close for comfort, WXTZ has complained.

As a practical matter, when a radio rating firm hires indifferent women to make telephone surveys, it's unlikely they'll ask, "Did you say, tee as in Tom and z as in Zebra, or cee as in Charlie and bee as in Baker?"

Burden had asked the FCC to give him the WXCB call letters by next Tuesday—the day he has to give up the WIFE-FM call letters as part of the transfer deal with the new owners of WIFE-AM, who will keep the WIFE initials.

THE INDIANAPOLIS STAR — WEDNESDAY, MAY 26, 1976

TV SCENE

New Owners Ready To Take Over WIFE

By JULIA INMAN

WITH A technical OK expected from the Federal Communications Commission this week, Indianapolis Broadcasting is ready to take over the long-disputed WIFE-AM June 1.

"We'll keep it a rocker," said Jerry L. Kunkel, who will be general manager of the station as well as executive vice president of the group of 11 local businessmen who will own the station.

The take-over will come on the heels of the U.S. Supreme Court's declining to hear the license renewal case for WIFE owner Don Burden's other stations including WIFE-FM here.

"DON HAS shed his bur-den," quipped an Indianapolis Star reporter, yesterday. Certainly it has been a long, fascinating battle over the Burden interests which should go down in the annals of broadcast history.

Indianapolis Broadcasting will keep the call letters WIFE-AM, Kunkel said, but Burden has applied for the letters WXCB for WIFE-FM.

Kunkel also announced that Al Kucin, formerly with WIBC here and with RKO General, will be general sales manager.

PROGRAM director will be Sam Holman, who has been with KQV at Pittsburgh, WLS at Chicago, where he originated the National Barn Dance, and WABC at New York, where he was program director at one time.

Kunkel's biggest coup, he feels, will be the annexing of Pittsburgh disc jockey Jack Armstrong, a legend in broadcast circles. Armstrong will have the 7 to midnight shift, but most of the station's other on-the-air personalities will remain the same.

Armstrong has been described as being able to press the "emotional start button" for listening teen-agers.

★

SPEAKING of call letters, WRTV (6) has changed its mind about reverting to the station's old letters WFBM-TV.

The outlet's application to the FCC was approved but the station took a survey and discovered that WRTV was firmly established in the public's mind so that no switch will be made.

★

SATURDAY'S "500" Festival Parade will be aired on 194 TV stations throughout the country, according to WRTV(6) General Manager Jerry Chapman.

The station's hour-long coverage will not be shown here until Sunday at 8 p.m.

bara 'Walters". Miss Walters is scheduled to leave NBC June 4.

She will appear then on a two-hour Today tribute to her native state of Massachusetts.

ABC News also revealed that Howard K. Smith, the veteran commnentator on the network's evening news programs, will be anchorman of its Close-up news specials starting in January of next year.

Mrs. Brown To Get Little Red Door Award

Mrs. Marvin Brown, 8002 Canary Lane, will be honored for 5,000 hours of volunteer service at the Little Red Door's annual Volunteer Recognition and Awards Day at 1:30 p.m. June 2 at the Booth Tarkington Civic Theater, 1200 West 38th Street

"We'll keep it a rocker," said Jerry Kunkel, the new general manager of WIFE, for an article in the *Indianapolis Star*. Bob Kiley stepped down as general manager to form an ad agency and Pat Kelly, who had been sales manager, filled the role.

WIFE Radio Going 'Adult Contemporary'

After 15 years as a youth-oriented "Top 40" radio station, WIFE is changing to an "adult contemporary" format.

WIFE general manager Pat Kelly said the station will make the change Monday.

"We are switching our target emphasis from teens to adults in the 25 to 54 age brackets," Kelly said. "Not only will WIFE Radio play exclusively adult-oriented music, but we will increase our emphasis on entertainment, information and sports."

The change became necessary be- 15 years ago," he said. In 1963 people who were about 20 are now 35 or 40. "We thought it was time to grow with them."

Another change will include broadcasting all Cincinnati Reds baseball games, starting at 7:05 p.m. Tuesday. WIFE will broadcast the remaining 42 Reds games this season and will continue total broadcast coverage next year.

Kelly said the station, 1310 on the dial, also is considering carrying other sports broadcasts.

WIFE's new owners did their best to maintain the exciting and vibrant brand they now controlled. They continued with Top 40 music for a while but changed to adult contemporary and eventually even tried talk radio.

The station limped along, tried different formats, but eventually the WIFE call letters were changed. And it was gone. —Al Stone

THURSDAY, JANUARY 6, 1983 — THE INDIAN

TV Scene *by Julia Inman*

Staff fired at WIFE-Radio

Sportscaster Paul Scheuring was planning on taking a one-week honeymoon next week, but events Wednesday canceled both his honeymoon plans and his job.

Scheuring and 12 other WIFE Radio employes were let go in what was called an "extensive" cutback

TV & RADIO

we don't have the money to keep them," Mueller said, citing the fact that the station is operating under Chapter 11 bankruptcy.

on the air with some minor changes," Mueller continued.

"I think the format is successful in this market and we've had great listener response. There is every indication that the ratings will improve."

Seven

The Buzzard Falls

Towards the end of the 1970s, Fairbanks Broadcasting had the no. 1 station in town, WIBC, as well as the no. 2 station, WNAP. The Hilliard/Johns strategy had worked perfectly. Unfortunately, it was too good of a plan, as WNAP was a victim of its own success and came to an end.

WIBC was making so much money that Jim Hilliard was able to buy new stations for the company with cash. Since there was no debt load, George Johns was given that amount as a promotional budget for the new acquisition with the mantra: protect WIBC at all costs.

From its inception, the purpose of WNAP was to be a "gunboat" that would protect the mother ship, WIBC. If a threat was detected, WNAP moved through the waters to dispatch the problem by adjusting its programming. It needed to be adaptable. When WNAP became the second-biggest station in town, it was not able to respond and itself became a threat to WIBC.

As the 1970s progressed, there was a strong opinion at Fairbanks that AM was doomed and FM was the future. Soon, the thought was that WIBC, the AM station, would sink in the ratings, taking all of the revenue with it. So, wheels began turning to slowly change WNAP into an FM version of WIBC. The idea was that when the day came, WIBC would switch to FM and the audience would already be there. The music at WNAP turned to adult contemporary, and an older audience was cultivated. But, as the 1980s came in, WIBC never lost steam; it remained a ratings winner and powerhouse for Fairbanks Broadcasting. The market, though, did not need two flavors of WIBC.

In the meantime, Richard M. Fairbanks sold his stations. The studios changed and so did management.

You can't make that many changes. When you start messing with something, the people will go away, and it's real hard to get them back. —Ann Craig-Cinnamon

They tried to be all things to all people at one point, and that never works. You know, the audience gets tired of the see-sawing back and forth. They want what they want. They can find—and they were finding other alternatives to WNAP at the time. —Pam Carr

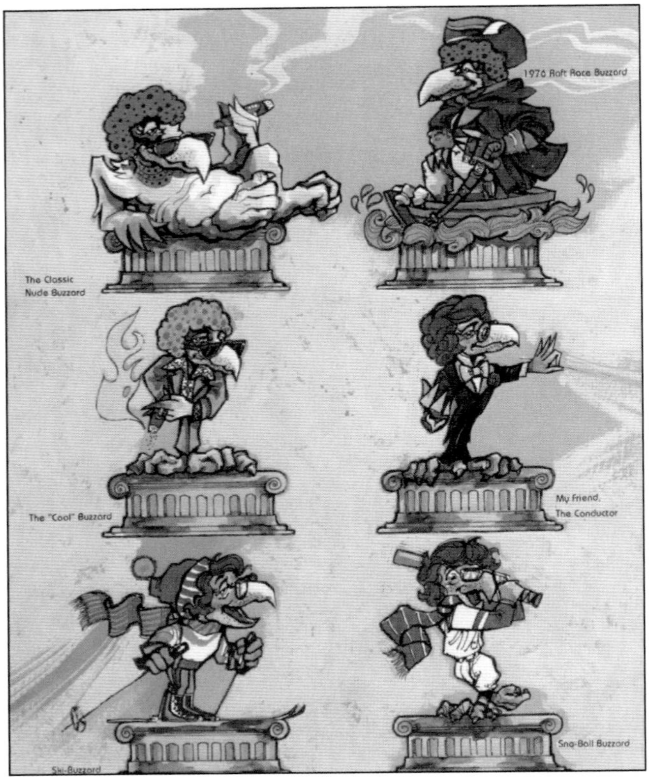

Perhaps the commercialization of the station's buzzard mascot serves as the best illustration of the station's swift decline. What had once been a symbol of antiestablishment rebellion was being dressed up in leisure suits, smoking cigars, and skiing the slopes. What had happened to the "bad boys" who sounded like they were going to swear any second? The station became as polyester as the suits its account executives wore.

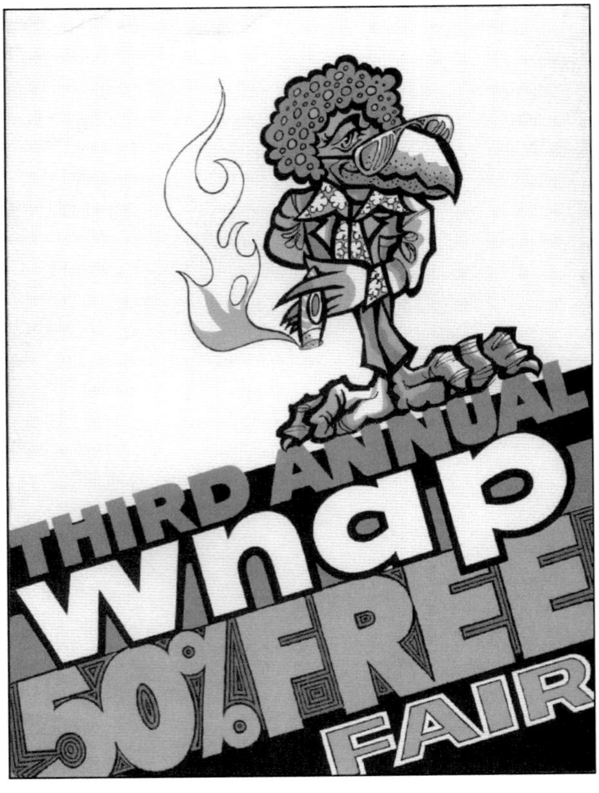

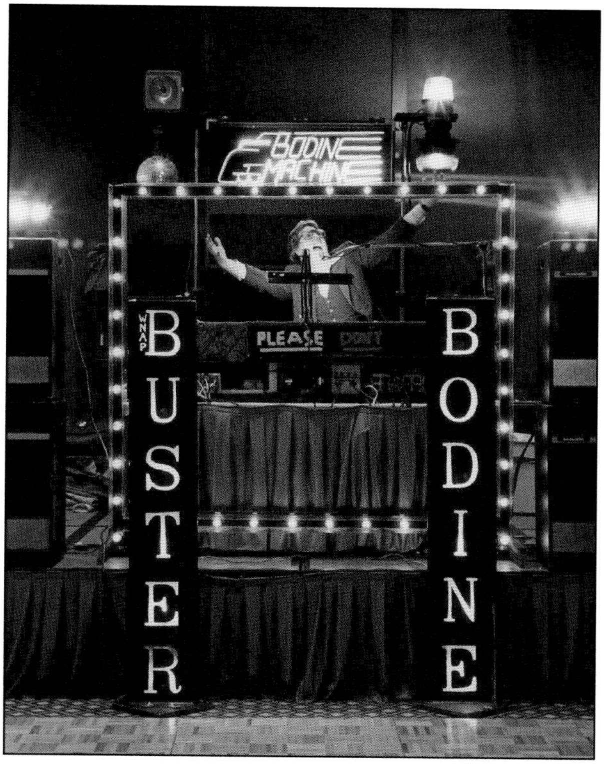

Michael D. Hanks put together a remarkable traveling show that was available for hire. The "Bodine Machine" featured a world-class sound system and flashing disco lights. It epitomized WNAP's success.

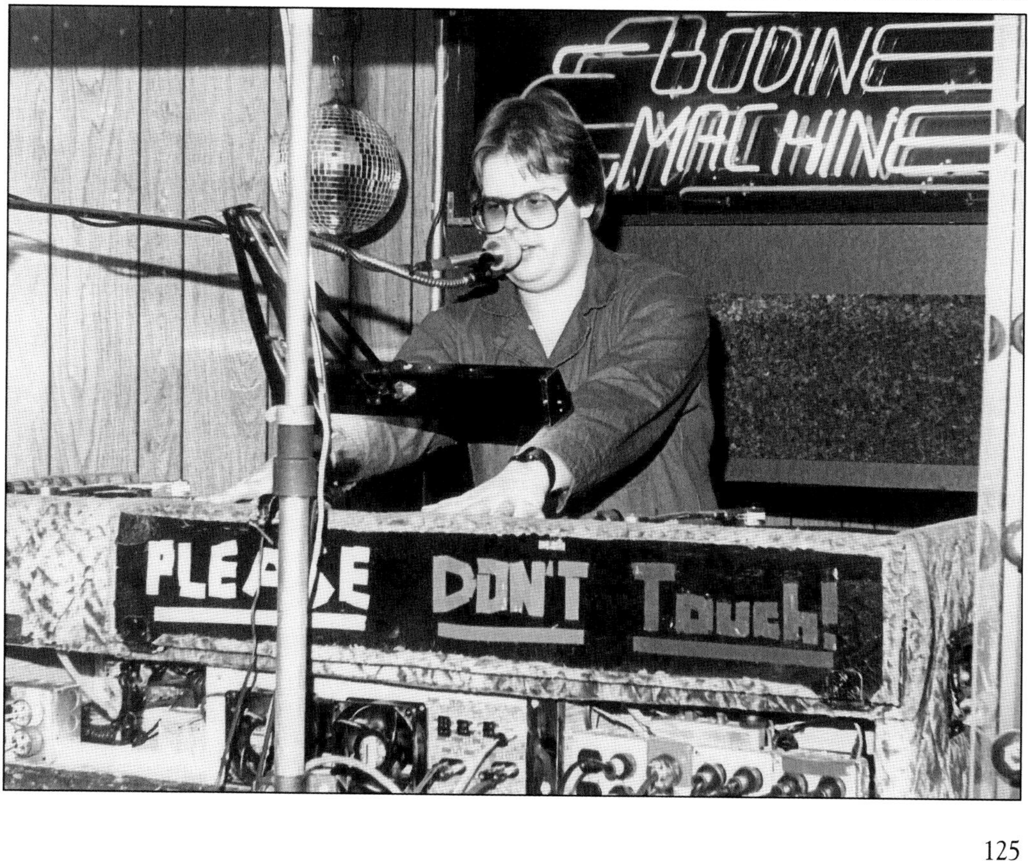

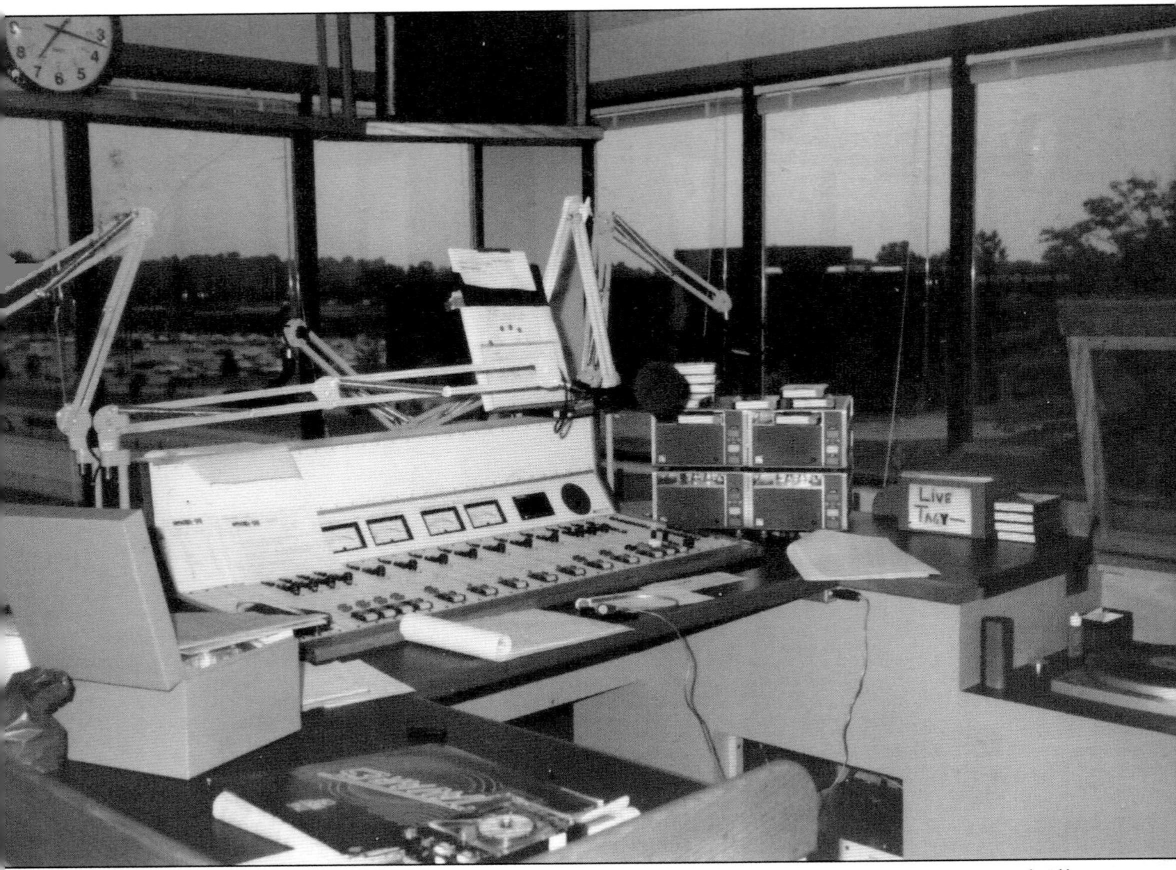

In the early 1980s, Fairbanks Broadcasting moved out of the "Two-Story Brick" on North Illinois Street and into a brand-new building on North Meridian Street that was clean, sterile, and modern. It was everything the original studio, in a closet at the back of the old building, was not. Many believe the new studio killed "the dangerous and dirty beauty" of the station, and in turn killed WNAP.

A PRODUCTION OF VIDEOPOLIS.TV

IN ASSOCIATION WITH KING MEDIA SERVICES

A FILM BY DAVID FULTON

NAPTOWN ROCK RADIO WARS

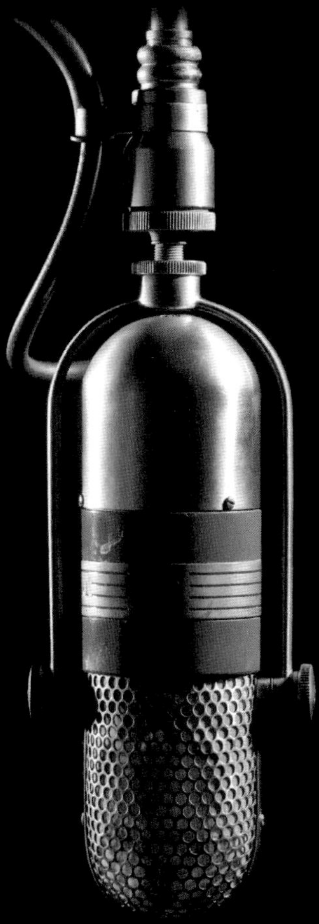

It was a rock 'n' roll death match.

Listeners had the ringside seats.

A VIDEOPOLIS.TV PRODUCTION IN ASSOCIATION WITH KING MEDIA SERVICES PRESENTS A FILM BY DAVID FULTON "NAPTOWN ROCK RADIO WARS"
BUSTER BODINE CRIS CONNER ANN CRAIG BILL DONNELLA TOM COCHRUN STEVE COOPER FAST FREDDIE FEVER HAL FRYAR 'BIG JOHN' GILLIS MIKE GRIFFIN
'SHOTGUN TOM' KELLY JIMMY MAC ROGER W. MORGAN MIKE 'THE ANIMAL' O'BRIEN REB PORTER SCOTT SHANNON BILL SHIRK JEFF SMULYAN AL STONE
ART VUOLO AND MORE EXECUTIVE PRODUCER DAVID FULTON EDITOR BRAD SCHUCHARD PRODUCED BY DAVID FULTON BRAD SCHUCHARD AL STONE DIRECTED BY DAVID FULTON

THIS FILM IS UNRATED VIDEOPOLIS.tv

WWW.NAPTOWNROCKRADIOWARS.COM

Consistent with our mission to preserve history on a local level, this book was printed in South Carolina on American-made paper and manufactured entirely in the United States. Products carrying the accredited Forest Stewardship Council (FSC) label are printed on 100 percent FSC-certified paper.

MADE IN THE USA